T0205506

Springer Biographies

The books published in the Springer Biographies tell of the life and work of scholars, innovators, and pioneers in all fields of learning and throughout the ages. Prominent scientists and philosophers will feature, but so too will lesser known personalities whose significant contributions deserve greater recognition and whose remarkable life stories will stir and motivate readers. Authored by historians and other academic writers, the volumes describe and analyse the main achievements of their subjects in manner accessible to nonspecialists, interweaving these with salient aspects of the protagonists' personal lives. Autobiographies and memoirs also fall into the scope of the series.

More information about this series at http://www.springer.com/series/13617

Wolfgang Osterhage

Johannes Kepler

The Order of Things

Wolfgang Osterhage
Wachtberg, Nordrhein-Westfalen, Germany

ISSN 2365-0613 ISSN 2365-0621 (electronic)
Springer Biographies
ISBN 978-3-030-46860-6 ISBN 978-3-030-46858-3 (eBook)
https://doi.org/10.1007/978-3-030-46858-3

© Springer Nature Switzerland AG 2020
This work is subject to copyright. All rights are reserved by the Publisher, whether the whole or part of the material is concerned, specifically the rights of translation, reprinting, reuse of illustrations, recitation, broadcasting, reproduction on microfilms or in any other physical way, and transmission or information storage and retrieval, electronic adaptation, computer software, or by similar or dissimilar methodology now known or hereafter developed.
The use of general descriptive names, registered names, trademarks, service marks, etc. in this publication does not imply, even in the absence of a specific statement, that such names are exempt from the relevant protective laws and regulations and therefore free for general use.
The publisher, the authors and the editors are safe to assume that the advice and information in this book are believed to be true and accurate at the date of publication. Neither the publisher nor the authors or the editors give a warranty, expressed or implied, with respect to the material contained herein or for any errors or omissions that may have been made. The publisher remains neutral with regard to jurisdictional claims in published maps and institutional affiliations.

This Springer imprint is published by the registered company Springer Nature Switzerland AG
The registered company address is: Gewerbestrasse 11, 6330 Cham, Switzerland

Preface

There exist a number of good books about Johannes Kepler. Some are purely biographical with emphasis on historical facts embedded in the important events occurring in Kepler's lifetime. Others investigate the theological soul-searching of the astronomer, and still others are interested purely in his scientific achievements. This book puts the person into another perspective, as the subtitle "The Order of Things" suggests.

When we consider Kepler's life achievements and intentions, the one leitmotif that drove his endeavors from his youth to his very end was the quest for order—or rather, the quest to discover the unique order underlying all things. Although he may have been the most thorough of natural scientists engaged in this quest, he was by no means alone on this path. Why is that so? What are the underlying reasons why people would follow such a course? This raises the question as to why science is practised in the first place. In particular, what is the origin of scientific thinking and observation? The search for order in chaos came a long way and reached its culmination in the cosmic harmony attempted by Johannes Kepler. This is what this book is all about.

After a brief introduction, the book starts with some considerations about what drives people in general and what drove Kepler in particular to seek underlying order in the cosmos, followed by a broad description of the major historical events during Kepler's lifetime. It then proceeds along a biographical route using his places of residence along this winding road as milestones. His major achievements are discussed and put into perspective by referring them to ancient and modern concepts.

At the end, a comprehensive timeline presents the most important historical and scientific events, including those of Kepler's own life. My thanks go to Springer International Publishing for making this book possible in the first place. Special thanks are due to Dr. Angela Lahee and her team for their support.

Wachtberg, Germany | Wolfgang Osterhage

Contents

Chapter 1
Introduction

The German physicist Carl Friedrich von Weizsaecker recognized Johannes Kepler as a genius in his book "Grosse Physiker (Great Physicists)" [1] on the basis of the following considerations.

One could contend that the true revolutionary discovery at the onset of modern theoretical astronomy was not the Copernican system, but Kepler's 1st Law, which tells us that planets travel on elliptical rather than circular orbits. Kepler succeeded in formulating it using the painstaking observations made by Tycho Brahe. In von Weizsaecker's view, it was a rare stroke of luck in the history of the natural sciences that Tycho's voluminous collection of data ended up in the hands of a scientific genius like Kepler, who possessed both a creative imagination and an affinity for scrupulous exactitude. Kepler believed more than anyone else in the mathematical perfection of celestial spheres. That is why he initially refused to accept the aberration of 8 min between the observed and calculated motions of Mars. He finally relinquished the circular path after trying more than 40 different orbits of Mars, none of which matched with observation. There remained only the ellipse, and this shattered his world view. But even then he remained confident that ellipses could nevertheless be elements of a harmonious world model, just like circles.

In Kepler's own understanding, the process of perception involved mapping the perceived onto his inner ideas and then assessing their concordance. Because just as external encounters remind us of what we knew beforehand, sensory experiences—once recognized as such—elicit intellectual and internally present facts, in such a way that they then light up in our soul, whereas they were previously hidden, existing only as a potential. No ideas are received through discourse; all already existed beforehand.

For Johannes Kepler, the quantitative aspects of his three laws of astronomy were of minor importance to him. His insight into the motions of the planets served to discover the plan of creation and finally to glorify the Creator. This resulted in his new astronomy, and indeed a new science. After Kepler, no one would ever integrate geometry, music, astronomy, and theology the way he did, although it remained difficult for anybody else to understand at the time.

© Springer Nature Switzerland AG 2020

W. Osterhage, *Johannes Kepler*, Springer Biographies,
https://doi.org/10.1007/978-3-030-46858-3_1

This Book

A biography is of course in some respects a tale of the succession of notable events during the lifetime of a person. If this were the only interest, these events could be compressed into a timeline fitting on one page. A comprehensive timeline of this kind can be found at the end of the book. However, the book has as subtitle: "The order of Things", which emphasizes its main objective. In this context, the life and achievements of Johannes Kepler can be regarded as a kind of catalyst to contemplate humankind's age old endeavor to discover some sort of order in the environmental and cosmic events it is witness to. Kepler appears to be that person in history and science who carried this task to its extremes. The enormous volume of his publications is testimony to this.

The next chapter is a kind of prolog to this daunting task, going into some detail about the driving forces in the quest for world harmony and the role Kepler played in this. It is followed by a depiction of the historical, political, and economic context in Central Europe, which dominated the wanderings of the famous scientist. The rest of the book then follows Kepler on these wanderings through the various geographical locations he was more or less forced to settle in during his life, starting from infancy, when his restless parents were moving around Wurttemberg. Despite his situation at home, he finished his studies and exams in southern Germany, enjoying his first encounters with Greek mysticism and the work of Copernicus. His first professional position was in Graz, where he obtained a professorship, published his "Mysterium Cosmographicum", and corresponded with Galileo. Then he was forced to relocate to Prague, where he first worked with, then fell into disagreement with Tycho Brahe. Here he had the opportunity to have a look at the Mars data, and the Tabulae Rudolphinae, which he completed later. In his "Astronomia Nova", he published his first and second laws of planetary motion, produced several more publications, and improved the telescope.

After leaving Prague, Kepler went to Linz. He was torn between Lutheranism, Calvinism, and Catholicism. He published his magnum opus, the "Harmonices Mundi", in which he expounded his third law of planetary motion and developed his own logarithms. After completing the Tabulae Rudolphinae, he went to Ulm, where he stayed for a short time. He then went back to Prague where he met Wallenstein, who asked him to work out his horoscope. His final occupation was in Sagan, as court astrologer, before he died in Regensburg.

In the last chapter, we revisit "The Order of Things". Kepler's harmony was dismantled and deconstructed by his successors, but science later returned to reconstruct it in a different way through the unification attempts of modern physics.

Chapter 2
The Order of Things

Some years ago I gave a lecture entitled "Ordnung und Chaos in den Naturwissenschaften am Beispiel der Physikgeschichte" (Order and Chaos in the Natural Sciences Using the Example of the History of Physics). I went through the main cosmological protagonists from Anaximander to Eratosthenes, Aristarchus, Aristotle, and Ptolemy, then through Copernicus, Galileo and, of course, Johannes Kepler, to Newton. At one stage a student approached me and asked my opinion as to whom I would regard as the most important of them with regard to the subject I was teaching. She proposed that it would certainly be Kepler. I answered with a nod.

When I reflected on the question later, I concluded that it did not make sense as it stood, but could only have some sort of meaning with respect to an agreed yardstick, for example, with respect to other achievements in astronomy, physics, or cosmology, or again in the quest for a harmonious depiction of the world—a world model comprising many features of observation. Going by the latter, Kepler must certainly be counted as one of the greatest.

Johannes Kepler was not actively seeking any disagreement with antiquity, but his was the final attempt to fathom the secrets of a supposed world harmony. Harmony thus became another issue to bother the scientific mind. In particular, Kepler was looking for a different reality behind the things which he tried to decipher by rational means. Furthermore, throughout his life, he was involved in a second discipline relating to the stars—astrology, which he did not regard as being in any contradiction with astronomy as such. This is what allowed him to remain a mainstream figure in his own time.

In contrast to Galileo and Copernicus, who maintained the assumption of circular planetary orbits for esthetic reasons, Kepler developed distinct laws based on detailed observations by himself and Tycho Brahe, leading to elliptical paths. Against this backdrop, Kepler was looking for commensurabilities of a special kind. His magnum opus dealt not so much with mathematical and physical formulas as with the search for a perfect harmony, expressed by planetary orbits, the harmony of musical notes and the relationship of geometrical bodies between themselves. Kepler believed he had proved these harmonies, while employing his numbers in quite a generous way.

© Springer Nature Switzerland AG 2020

W. Osterhage, *Johannes Kepler*, Springer Biographies,
https://doi.org/10.1007/978-3-030-46858-3_2

Fig. 2.1 Henri Poincaré

When approaching Kepler's life achievements and intentions, the one theme that drove his endeavors from his youth to his very end was the quest for order—or rather, the quest to discover the one singular order underlying all things. Although he may have been the most thorough of natural scientists in this quest, he was by no means alone on this path. Why is that so? What are the underlying reasons for people to follow this course? Here are some considerations.

The French mathematician Henri Poincare (1854–1912, Fig. 2.1) once claimed: "We are fortunate to be born into a world in which events happen which recur. Imagine if we had to deal with 80 million chemical elements instead of the eighty we know, and that not only some of those were commonplace and others rare, but that all were evenly distributed. This would mean that every time we picked up any stone there would be a high probability that this stone would be composed of an unknown substance. In such a world there would be no science at all. But thanks to providence this is not the case."

However, one could argue that even in such a world there would be science. People would find some way of systematizing things, just as has been done in botany, for example, to reduce the amount and variety of foliage of trees to some manageable classification scheme. In this context the views of another Frenchman, the post-structuralist Michel Foucault, who lived and worked somewhat later (1926–1984), are of some interest. In his book "Les Mots et les Choses" (The Order of Things) [2], he got to the bottom of this problem. He claimed that the objective of any science was indeed not to describe objective insights but rather to create abstract concepts which were generally discussable. In that case, what we refer to as scientific progress would just be a continuing alternation of forms that would mutate throughout the course of history. Arguably, this may be so. On the other hand this process would mutate history itself! However, in Kepler's opinion the description of reality coincided with the description of truth itself. His claim was absolute, with not the least bit of room left for relativism.

In this context the question arises as to why science is practised in the first place. In particular, what is the origin of scientific reporting and reflection? Of course, this question has been pondered by generations of philosophers and historians. It will not be solved here. But there is a related aspect concerning this problem: what are the additional motivations for scientific endeavor, besides the pure quest for insight? Why were certain efforts exerted in just those directions? Or to put it another way, why are people reluctant to just take 80 million elements for granted and leave it at that?

One of the major drives for scientific research seems to have been to detect some sort of order in the chaos man came across. This assumes that there was a subjective perception of chaos in the first place. To avoid being engulfed by it, the need arose to describe it in such a way that it would look like order, since the discovered state itself would not of course change by being described as such: this was the formulation of order as an instrument to co-exist with chaos and hence domesticate it.

The ordering of chaos came a long way, and reached its culmination, in the cosmic harmony attempted by Johannes Kepler. Whatever other, underlying driving forces that may also have existed for him, he derived his motivation from his religion and from his personal beliefs, which did not fit the formal classification of the confessions of his day. He needed no other pretext at all. After Kepler, and especially in modern times, science has descended from this high peak by creating concepts that negate absolute space and absolute time and replace certainty by probability. And physics has still not reached any sort of final point today, so further efforts are being made to develop new unified theories that will take us from this intermediate situation and allow us to order things again. This occurs again and again to evade Poincare's dreaded state, always trying to reduce the famous 80 million elements to as few as possible. The mastery of chaos is thus "only" a semantic exercise.

We could continue to contemplate further on a purely philosophical level, since such considerations are relevant to the whole of science, but we will limit the perspective of this book to physics.

The first question raised when man turned his eyes to what was visible above his head may well have been: how can these phenomena be understood in space and time? The obvious next step would be to assume that behind these phenomena there is some sort of system, which could perhaps be deciphered, if only one could seek advice from the quantities governing them. Before taking aim at the real causes, scholars were satisfied for centuries just to describe the things they saw, and that was that [3]. And initially, that was also Kepler's sole aim. He contented himself with the beauty of harmony as such, without looking for the underlying causes—unless they were to be found in God's plan itself.

In spite of this there always existed a desire to find relationships which were as simple as possible and which could be used to make predictions, for example. Max Born wrote to Albert Einstein in 1948: "Concerning simplicity, opinions vary. Is Einstein's law of gravitation simpler than Newton's? Professional mathematicians would say 'yes'; and would refer to the logical simplicity of its foundations. Others would explicitly refute that because of the nasty complexity of the algorithm." Einstein responded: "It only depends on the logical simplicity of the foundations" [4].

The quest for simplicity implied a search for commensurabilities, i.e., to find—if possible—preferably whole number ratios between planetary orbits. These commensurabilities where thought to be the key to cosmic harmony, including the music of the spheres already contemplated by the Pythagoreans. To arrive at this end, Kepler did not shun any effort, no matter how tedious. As we shall see, he went through 120 permutations of possible combinations of the Platonic geometrical bodies to find his solution—and naturally all the calculations were carried out by hand, without the help of any computing device.

In spite of the many relativizing observations in the course of the history of physics, it seems that the cosmos exhibits a number of very exact harmonies (and some of these still elude explanation by cosmological models). Examples are the linked rotation of many moons with close orbits around their host planets, i.e., a 1:1 spin-orbit resonance. In this kind of rotation, the rotational period is linked to its period of rotation. As a result the moon will always show the same side to its mother planet, which is the case with our own moon. Concerning the orbital periods of planets around the Sun, the exact value of the 2:3 spin-orbit resonance of Mercury attracts attention. The same ratio exists between the orbital periods of Neptune and Pluto. Another example that has not as yet been explained is the relationship between the orbital periods of Saturn's moons and the corresponding gaps in Saturn's rings; neither has Kepler's relationship between the succession of planetary orbits and the nesting of Platonic bodies—his commensurabilities.

However, there exists no commensurability in the calculation of the dates for Easter, which are surprisingly not based on any cabalistic interpretation of the scriptures with respect to cosmology, but rather on a transcription of the Jewish lunisolar calendar to the Julian one, taking into account the historical tradition of the death and resurrection of Christ. This was fixed (on a theological basis) during the Council of Nicaea in 325: Easter should always be celebrated on the first Sunday after the first full moon after the spring equinox. There are two astronomical parameters to be considered:

- the date of the full moon, and
- the spring equinox.

Another example is the seven day week, which has no astronomical basis. There exists a long history of more and more refined algorithms to find a solution to this problem, which we shall not discuss any further here. It culminated in a procedure by Carl Friedrich Gauss with a correction algorithm, dating to 1816. In short, the development of numerical methods is a by-product of astronomical computation ever since Kepler, and this includes the invention of logarithms.

Another activity connected with astronomy was the calculation of the future by examining the positions of celestial bodies and thus somehow divining the will of the Gods—a phenomenon which continues to this day as astrology. This age old desire of mankind has still not been satisfied, and indeed it lingers on. For Kepler, astrology was an additional source of income. He was already famous as a student for his "prognostica". The most famous of these was the horoscope for Wallenstein (Chaps. 8 and 9).

It seems that, apart from the natural desire for insights into the functioning of nature, there are other sources of motivation in the search for order. Maybe in the end it is these that consolidate this desire and produce tangible results somewhere along the way. They are:

- Esthetics
- Homeostasis
- Gestalt
- Mythology
- Communication
- Power
- Benefit.

Esthetics

Recently, there appeared a newspaper report about a rubbish dump near Frankfort which had won an award for being the most beautiful rubbish dump in Hessia in Germany. This was not because of the arrangement of the items of rubbish, but referred to the design of the access road and the general structure, roofing, and commercial buildings. The arrangement of rubbish in itself would not naturally lead to considerations on harmony. We all know this from road building sites, when slabs of concrete from old paving have been piled up at the roadside. Instinctively, we know that such an arrangement has nothing to do with order. It represents "disorder". On the other hand, we immediately recognize when something is well ordered: a well tended park or a store house in which everything is at its rightful place on the shelves.

An impelling motive for scientific description is to find an algorithm which is as esthetic as possible, and which not only fulfills its scientific purpose, but represents at the same time some sort of beauty. For instance, when the mathematician Grigori J. Perelmann proved the Poincaré postulation about closed surfaces in a topological manifold some years ago, people said this derivation was "beautiful", and thus admirable [5]. In addition, the fact that it did not take a hundred pages, but only a few, also carried some weight. Hence, beautiful also rhymes with short. Another example was the comment that Enrico Fermi made to his assistant Ettore Majorana, who later went missing, when Majorana showed him his theory of the positron—a theory which concurred with Dirac's. Fermi said that Majorana's theory was worth publishing if only because it had been derived so wonderfully, as could be seen by merely glancing at the paper. In the end, Majorana declined to publish it [6]. And finally, I once heard a professor lecturing on the design of cars who claimed that, once an engine part met all esthetic criteria, it would surely also meet technical ones.

Kepler's esthetic appraisal of the "Mysterium Cosmographicum", his first major work, culminated in a final psalm-like hymn in praise of the Creator, exemplified by the following characteristic verses:

> ....
>
> But I look for the traces of Thy spirit out in the Universe,
>
> I regard ecstatically the glory of the mighty celestial edifice,
>
> This artful work, the mighty wonder of Thy almightiness.
>
> I see how Thou hath determined the orbits according to fivefold norms,
>
> With the Sun in the middle to donate life and light,
>
> I see that her laws regulate the course of the stars,
>
> How the Moon achieves her transitions, suffers eclipses,
>
> How Thou scatterest millions of stars across the realm of the skies.
>
> .... [7]

The principle of equilibrium or balance is very similar. In the visual arts, the principle of symmetry or weighted equilibrium played a major role from early on. This was only given up in modern times by alienation. Undisturbed symmetry creates a feeling of relaxation and is associated with harmony. The meaning of symmetry in modern physics has been elevated to an almost axiomatic principle. It plays an important role as a pillar in the Standard Model of elementary particles. Symmetry-breaking is one of the major crimes a physical experiment may permit itself. It will immediately lead to the introduction of new parameters in the model, to continue to preserve symmetry. In the search for a Grand Unified Theory (GUT), the symmetry between the hadrons and leptons plays a fundamental role.

Homeostasis

The concept of homeostasis points in a similar direction. Homeostasis describes a model according to which man is ever searching for optimal conditions that will create a certain equilibrium between internal and external processes, an equilibrium which one would like to preserve or restore in case of aberrations. It took Johannes Kepler a lot of soul-searching before he finally accepted, while grinding his teeth, that the planetary orbits do not follow the ideal figure of a circle, but are in fact elliptical. The circle in antiquity, in Galileo and Kepler's time, and even today is regarded as the most perfect of geometrical figures.

In the context of gaining scientific insight, we are not only considering life processes, but also the description und understanding of the environment in which life takes place.

When there are deviations, it is important to steer things back in the same way as one would direct a ship back on course when it is hit by a storm or carried along by a current. A modern example is the introduction of colors for quarks, which were introduced to allow a further degree of freedom, since otherwise the Pauli principle would have been put at risk. Here is the background.

Quarks in the hyperon states of baryons have to have an orbital angular momentum of zero for reasons of energy balance. But the spin and wave functions are symmetric under permutation of identical quarks. Thus the overall wave function is also symmetric, because the orbital angular momentum is zero. This is not allowed, since quarks are fermions with spin ½ and should therefore have an asymmetric total wave function. To solve this problem, quarks are assigned an additional degree of freedom which can assume three different values called colors: red, green, and blue. Now we have 6×3 quark states instead of only 6, since every quark can appear in three colors [8].

Once again, the above-mentioned symmetry relationships play a decisive role only because, in order to preserve them, counter-steering was necessary. There is no reason to believe that the cause for the further development of high energy physics was homeostasis. The real reason for introducing color into the quark model can certainly be found in the fact that the theory has so far, by and large, been able to describe the associated observations in an excellent way. In the end, the whole theory of quantum chromodynamics was deduced from quark colors to form the basis for the quark model that is generally accepted today.

Gestalt

When looking for motivations to systematize insights in the natural sciences, especially in physics, it is useful to consider the impact of Gestalt psychology.

Empirical investigations have led to the conclusion that certain compositions of forms create different stimulus reactions in observers, when they look at them. These reactions depend on the visual nature of such pictures, especially when they contain abstract forms. Similar observations can be made on the basis of sounds and chords. Other results of these studies indicated that there is a high degree of agreement between different observers when they evaluate their perceptions. One example is a right-angle in a figure when compared with an angle deviating by 10° in a different figure. While in the first example a certain harmony was perceived in the overall context, the deviation from it was regarded as a perturbation. The same goes for lines that cut across one another, etc. And we have already mentioned the circle and Johannes Kepler's problem with ellipses.

Gestalt psychology has contributed to the explanation of such reactions, which cannot be explained on a rational basis (apart from practical considerations, why should the rooms in a house always have perfectly vertical walls?). In this context, Gestalt refers to a quality attributed to an observation. To put it simply, something has Gestalt if it is rich in content, if it makes sense. Accordingly, there is apparently a "good" Gestalt to be pursued—again, in the interests of homeostasis and balance. A "bad" Gestalt, leading to a feeling of unease experienced by the observer, exerts the drive to correct something. Since there is a high rate of agreement between individuals with respect to a good Gestalt, it is assumed that Gestalt perception is a hereditary reflex. This can be traced back to the human ability to compose a meaningful whole from fragmented observations. The training of artificial neural networks for pattern recognition shows that, in spite of this, an element of practice is still necessary.

As already pointed out, the beauty of a mathematical equation or a mechanical construct may provoke a reaction according to Gestalt psychology. It remains to be seen what influence the Gestalt of geometrical objects may have had a priori, up to the Riemannian space-time continuum of the general theory of relativity.

Mythology

Going back to the origins of ancient cosmologies, one inevitably comes across mythological attempts at explanations of genesis and meaning. It is useful at this point to explain some basic differences:

a. Mythological world models do not claim to be scientific in the sense of present day natural science, but contain theological propositions originating from different preconditions.
b. These mythological world models are not necessarily preconditions for the later development of rational models.

Certainly, religious and poetic attempts have been made to offer orientation to human beings in his world environment [9]. In this context, one speaks of a cosmogonic-theogonic schema: from an undefined original state, the present world of Gods, humans, the heavens, the sea, and the Earth originate from a sequence of generations of individual divine powers. Literature and mythology may have contributed to the development of philosophical attitudes. On the basis of known records, the pre-Socratic philosophers were the first to develop cosmological theories outside a mythological context. And one of these who presented a concise model for the creation of the world was Anaximander, who lived from around 610 to around 550 BC. Although no original document from his own hand has been preserved, philosophers and historians who came after him have handed down his findings to subsequent generations. The development of today's natural science, however, involves observation of phenomena and their relationships to one another as a prime condition.

In this context, we should make some remarks about the oft-cited history of the creation in the Old Testament. This was not intended as a competing mythological schema for other creation myths, but rather as an enlightened scripture given the conditions prevailing at that time [10].

Common understanding imagines a nomadic shepherd sitting in an oasis with his small livestock (that is, goats and sheep), expressing his admiration for the beauty of the world during a long period of rest, just as some passages of the Genesis describe it. In this respect, one could claim Genesis to be a somewhat poetical opus. However, one has to bear in mind that the final editing of this book took place during the exile of the Jewish upper class in Babylon. The prisoners were impressed by the processions during heathen festivals and the presentation of Babylonian Gods, and unsettled by the power of their presence. To counter this impression, they put together their own view of God and the world, starting with the 1st Book of Moses. In this way the creation account was intended to determine a political position, de-mythologizing the world of beliefs of the Babylonians: the stars were not Gods, just lamps made to measure time. This was the first step toward the de-divinization of nature, which in other cultures was indeed populated by all sorts of Gods and spirits.

On the other hand this model did not serve as a basis for rational advancement toward genuine scientific theories. It remained what it was intended to be: a theological basis for monotheism, but without any pretense of being a scientific or historical document. On the other hand, there is no doubt about the fact that Johannes Kepler firmly believed that the whole world had been created on the basis of a master plan made by God, and this included its mathematical foundations.

Communication

The already mentioned esthetic point of view applied when composing formalisms evidently recedes into the background given the basic need for abstract perceptions, so that nobody is forced to talk about all 80 million different elements to explain his world model. This meant that, for reasons of communication, a language had to be found to describe scientific issues in a short but precise way. This language was found in mathematics. The chicken-and-egg problem—in this case, which was first, mathematics or the formalization of issues resulting finally in mathematics—is really of no relevance. If one traces the development of calculating machines or aids to calculation well back into antiquity, it seems that mathematics owes itself more to a mercantile origin than a scientific one. It is true that mathematics and physics have always cross-fertilized each other and are still doing so today. Logarithms and computation in astronomy during the early stages of the modern era provide just one example. There has been a similar mutual influence between topology and the general theory of relativity or cosmology.

In this way, the need for communication of specific issues became a further precondition for structured descriptions in a scientific sense. When reading Nikolaus Copernicus, one gets an idea of how scientific language has changed since those times, precisely because translations seem to be required—and this not only concerning its specific features, but also regarding the technical description of observation methods. On the other hand, Kepler's vast body of publications (see the compilation at the end of the book) exhibits features which would be unthinkable in today's scientific papers: doxologies of God, articles of faith, and references to ancient mythologies mixed with geometrical considerations and observational data.

Today, there is a vast popularization of physics, while fewer and fewer people can actually trace back to the true foundations of the various issues. The reason for this is not to be found in the theoretical concepts underlying these issues, but rather in the formalisms developed for their description. In the course of the correspondence between Einstein and Born already mentioned, and which lasted for nearly forty years after all, they busily discussed the whole spectrum of physics: quantum physics, the theory of relativity, solid state physics, fluid mechanics, atomic physics, and so on, and indeed in depth. The situation today is that physicists only ever possess a limited understanding of the specific issues of colleagues involved in different areas of physics.

Power

As already mentioned, there is a very strong motive for the intellectual mastery of reality, in any way possible, and that is the desire to predict events. Anyone who could determine the will of the gods or the future behavior of the stars would not only add to the benefit of the community or the local ruler, but would also obtain a position of power himself. If only certain individuals possessed this ability, the others would end up in certain state of dependency with regard to these skills. This was certainly true for the early observations of the sky.

The question of power and influence, however, goes beyond astrological and astronomical objects. In modern times, scientists have exerted influence outside their specific topics of scientific interest. This can be illustrated by the influence of physicists on the outcome of World War II if we consider the problem of the atomic bomb, whereas during World War I, chemists still dominated the scene, because gas was being used as a weapon of war. After World War II, physics became the leading science for at least a generation, before it was replaced in this role by biological sciences such as genetics. The influence of the biological sciences can still be felt because of its sociopolitical relevance with regard to stem cell research and preimplantation diagnostics. But today one can witness another replacement process in the direction of neuroscience as the new leading science. The question of power is visible on two levels: one is the quest for social influence, while the other concerns competition for research funding.

In Kepler's case, the question of power first surfaced in his struggle with Tycho Brahe, who—as chief mathematician—engaged him as his assistant at the court of Emperor Rudolf II in Prague. To protect his own position, Brahe kept Kepler at the maximum possible distance from his own insights into the secrets of planetary data, making Kepler's life and work unbearable. After Brahe's death, the struggle continued with the inheritors of the data for the Tabulae Rudolphinae. And in the end, when Wallenstein, Kepler's last sponsor, fell from grace, Kepler himself remained without any material resources.

Benefit and Values

Closely related to influence and power was the question of benefits from scientific results. Besides demythologizing the heavens, one other important motive for looking at the stars was the desire to find a system that could harmonize recurrent planetary or stellar configurations with the rest of nature. One of the objectives was to deduce a harvest calendar that could be used to calculate dates for sowing, for example. The consequences of this early subdivision of the year can still be found in our own calendars today. The dating of Easter is one example. Christian Easter corresponds to the Jewish Pessach, which in turn follows from a date for sowing observed by the Canaanite natives before the conquest by the old Israelites.

The question of benefit is closely connected to another one—the question of value freedom. In 1904, Max Weber formulated his thesis about the value freedom of science. It states something like: no empirical science can lay down binding value input, although it can determine the consequences of certain value systems. This thesis has been interpreted to mean that scientific insights have nothing to do with values, and values have no place in the laboratory. A scientist should only try to find factual connections, since values always come with something subjective that would endanger the objectivity of science. Even in the sixteenth century, the English philosopher Francis Bacon required that scientific objectivity must do without subjective perspectives so that it can focus unrestrictedly on the subject of research. In this context two types of value are of interest:

1. Cognitive values, meaning values with reference to cognition. These determine science's claim to knowledge and thus the nature of science itself. Behind this lies the desire to find theories with a high degree of explanatory power, for example, using only a limited number of basic elements to describe a multitude of phenomena. Of course, if one required ten different hypotheses to explain ten phenomena, none of these hypotheses would probably be regarded as acceptable. One would always be inclined to assemble as few hypotheses as possible to explain a multitude of phenomena, in line with the already mentioned esthetic needs.
2. Finally—and this brings us back to benefit—utility values: scientific research contributes to the national economy.

However, quite often the search for truth is evoked as a driving force in contrast to benefit or utility. This is certainly true for the actual intentions in many cases, but only partly correct. In reality, the selection of research subjects takes place beforehand, since scientists would never be able to investigate everything. Thus there are truths which can immediately be classified as unimportant: nobody is interested in the actual number of grains of sand on a beach. This means that truth has to have an underlying meaning. Hence in practice, objectives are formulated that satisfy the requirements of economics and politics. The danger is that research which concentrates solely on the practical needs of everyday life will in the end stagnate with respect to basic insights. This was no option for Johannes Kepler. His quest for truth was bound to his endeavor to prove that the truth he deciphered from his astronomical observations and the resulting harmony of the world corresponded at the same time to the structure of God's master plan, already present before creation itself.

Against the above-mentioned utilitarianism, there stands a science tackling fascinating tasks. Quite often these are research problems in connection with already resolved riddles that require further treatment [11].

We have already mentioned Anaximander. Later on, Plato wrote that the world had been created in such a way that human reason would be able to understand it. This world remains forever in its original state. It is a living creature equipped with a soul and reason. The Sun, the Moon, and certain stars came into being just to enable the measurement of time.

Aristotle also stated that there had never been any proof that the world had undergone any change. He assumed that the Earth was at the centre of the world and a sphere surrounded by other spheres containing the rest of the world. During his lifetime, the circumference of the Earth was calculated to an accuracy of 85% of today's value. This older value still served as the basis for Columbus' calculations in preparation for his voyages of discovery.

The Muslim philosopher Avicenna, who lived from 980 to 1037 A.D. stated that

- time is a measure of motion, and
- space exists only in the imagination of human consciousness and has to be regarded as an entity separate from matter.

Near the end of the Middle Ages, Nikolaus Cusanus (1401–1464) concluded that all parts of heaven including the Earth were in constant movement. Shortly afterwards, Copernicus and Galileo came out with their decision in favour of their heliocentric world model. At around the same time, Giordano Bruno theorised that the Universe must be full of uncountable suns and uncountable earths.

Enter Johannes Kepler. In addition to the motivations listed up to now, which might have influenced him unconsciously, he brought along his own powerful incentive, which was his own theology. The mapping of his theology onto his observations of nature is the key to his extraordinary scientific results: cosmic harmony or the order of things.

Chapter 3
Time and Space

1571–1630

Time and space are fundamental concepts in physics, but also in people's everyday lives. If one wants to narrate someone's biography and discuss his achievements, this can only be done by taking the relevant historical and geographical background into consideration. Time and space together form the framework against which the world line of the person rolls out: successions of personal events—important ones and not so important ones—embedded in the epochal circumstances of the world at large. Figure 3.1 shows the historical events which might have played a role in determining the path Johannes Kepler was to take and in influencing his mind and thought.

However, simple history aside, our main interest will be focused on those occasions where our astronomer encountered matters and people relevant to his scientific achievements during his life. These will form the rungs of ladder, along which we shall accompany him. But apart from these purely circumstantial considerations, it will be the designation of Kepler's role that lies at the centre of our reflections—against the backdrop of man's eternal quest to come to grips with the environment he is thrown into, by trying to order things in time and space so that life may eventually make sense.

Peers

Kepler was of course familiar with Copernicus' publication (Chap. 4), but the spiritual roots for the impetus that drove him reached deeper—down to Nicolas Cusanus, who lived from 1401 to 1464 and had developed a mystical geometry. Another foundation, which in the end turned out to be decisive for the practical success of Kepler's endeavors was the existence of the Uraniborg observatory, built for Tycho Brahe. The famous Tabulae Rudolphinae, named after emperor Rudolph II, an immeasurable treasure of data, had been collected over many years in this first and most important observatory, and indeed spiritual center for astronomical research, at the onset of the modern age. Kepler would eventually gain access to this data. Brahe himself

© Springer Nature Switzerland AG 2020

W. Osterhage, *Johannes Kepler*, Springer Biographies,
https://doi.org/10.1007/978-3-030-46858-3_3

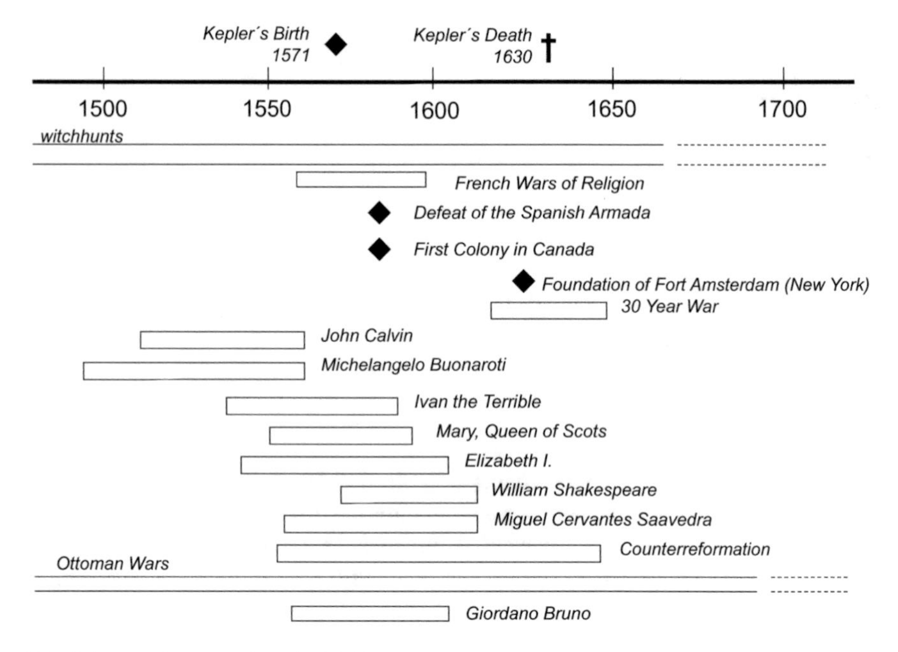

Fig. 3.1 Timeline around Kepler's Lifetime

accepted the Copernican world model only in part, and developed his own version, a compromise between the Ptolemaic and heliocentric systems.

The other great thinker of the day was of course Galileo Galilei, with whom Kepler communicated. This will be discussed in a separate section. In this context, Kepler benefited from the state of the art of the so-called Galilean telescope, which he later improved. Galileo had been born before Kepler and died when Kepler was already dead. Thus Galileo had been influenced by the same historical cataclysms as Kepler, although these events had different impacts on them, since the situation in Italy was quite different from that in war-torn Central Europe.

Events Shaping Kepler's Life and Thoughts

It was the time of the transition from the late Renaissance and the late Humanist period to the Baroque. Witch hunt s were on the rampage. The idea of witchcraft had already sprung up in ancient times. It was developed further and systematized during the Middle Ages to become a virulent source of social control. Witch trials came along in waves with periods of calm between them. There were peaks in the 15th and 16th centuries and again during the period that concerns us here, that is, during the lifetime of Johannes Kepler. The emphasis shifted from the accusation of just practising witchcraft as such by supernatural means to the accusation of direct association with the devil himself. The bulk of the witch hunts took place in Germany, although they did begin there later than in other European countries. Altogether witch hunts lasted for approximately 300 years all over Europe. During this period, about ten thousand witches, both male and female, were tried in southern Europe alone, and thousands of whom were actually executed, usually burned at the stake. It was Kepler's mother who was accused of witchcraft and subjected to a witch trial, but she was eventually released thanks to the intervention of her son (see Chap. 7).

Shortly after Kepler's birth, the French Wars of Religion between Catholics and Huguenots escalated after St. Bartholomew's Day in 1572. These wars refer to a total of eight different, successive wars over a time span of altogether 36 years. The belligerents were Protestants on one side, represented by the Huguenots, England, and Scotland, and Catholics on the other side, represented by the Catholic League, Spain, and the Duchy of Savoy. These wars were interspersed by various attempts to achieve peace, like the Peace of Longjumeau (1568), the Peace of Saint-Germain-on-Laye (1570), the Edict of Boulogne (1573), the Edict of Beaulieu (1576), the Treaty of Bergerac (1577), the Treaty of Fleix (1580), the Peace of Vervins, and the Edict of Nantes (1598). The latter brought some kind of end to the fighting, including some factional strife between different opponents in France itself. One of the unfortunate highlights of these conflicts was the massacre on St. Bartholomew's Day in 1572. The religious and political climate all over Europe was strongly influenced by these events.

The French Wars of Religion were by no means the only wars that took place during Kepler's lifetime. In the year 1588, a truly historic event occurred, which is still summarized to this day by just two words: the "Spanish Armada", actually short for the defeat of the Spanish Armada. This battle was part of the undeclared Anglo-Spanish war, which raged from 1585 to 1604. The belligerent parties involved were the Kingdom of England and the Dutch Republic on the one side and the Iberian Union with the House of Habsburg on the other.

The purported reason for this conflict was once again religious. However, it was in fact triggered by Henry VIII's desire to divorce his first wife, Catherine of Aragon. The divorce, however, was never granted by the Pope, and Henry VIII thus chose to break with the Catholic Church. A few years later, the English Reformation fell in line with the Reformation in other parts of Europe. This was followed by various attempts at Counter-Reformation s, but it was Elizabeth I who in the end firmly established Protestantism in England. King Phillip II of Spain, with the support of Pope Sixtus V, subsequently organized a crusade to win back England to Catholicism, and for this reason he assembled the Spanish Armada to invade England. As is well known, this attack resulted in complete failure. By a combination of battle engagements with the English fleet, commanded by Sir Francis Drake, navigational errors, and foul weather, the Spanish Armada was thoroughly defeated. Out of 130 Spanish ships 35 were lost and about 20,000 men died.

During Kepler's lifetime, the French began to colonize Canada. For most scholars, the beginning of the modern age coincides with the discovery of America in the year 1492—well before Kepler's birth. New horizons had opened up far beyond the geographical limits of Europe. This led to an enlargement of intellectual perspectives and had far-reaching consequences for human thought and reasoning about the world. During Johannes Kepler's lifetime, colonization was already well under way—notably in Canada, where it commenced with the founding of St. Johns in Newfoundland by Sir Humphrey Gilbert. This was the first English Colony on the American continent. It was later followed by the settlements at Port Royal and Quebec at the beginning of the seventeenth century. New France, as Canada was initially known, was rapidly colonized along the St. Lawrence River and the Great Lakes by missionaries and trappers.

New Amsterdam was founded in 1624 on the Hudson River. This followed in the wake of other renowned settlement activities, which had started further south in 1609: under Captain Hudson, the Dutch began to explore the area around what is known today as New York Bay. The drive behind this endeavour was purely commercial, based on the highly profitable beaver fur trade. A number of different expeditions followed in succession, resulting in the foundation of the New Netherlands in 1614. In the same year, the first European traders arrived in Manhattan. In 1624, the first permanent settlers arrived with their families, and one year later Fort Amsterdam was founded and the area around it was bought from a local Indian tribe in exchange for European goods representing a value of about 24 $ at that time—corresponding to 1100 $ in today's money—by Peter Minuit, the director of the Dutch West India Company in New Amsterdam, today New York.

Important Personalities

One of the most prominent figures of the Elizabethan era was the poet and playwright William Shakespeare, born in Stratford-upon-Avon in 1564. He died in 1616. His 38 plays are the most widely performed plays in the world, even to this day. He married early, at the age of 18, and had three children. He started his career as an actor, but soon began writing pieces for his own company of actors. His repertoire included comedy, historical drama, and tragedy. His works were widely published and distributed even during his own lifetime.

Another famous writer of that time was of Spanish origin: Miguel des Cervantes Saavedra, author of Don Quixote, born in 1547. He spent part of his early life in exile in Rome, between 1569 and1571, and thereafter, between 1575 and 1580, in the captivity of Ottoman pirates, who took him prisoner after he joined the Spanish navy during operations in the Mediterranean. Don Quixote was published with immediate success in 1605. Cervantes died in Madrid in 1616.

The Holy Roman Empire

The Holy Roman Empire served as a kind of umbrella organization for many territories in the middle of Europe, and provided a legal framework to maintain the coexistence between different sovereigns. These principalities were only partly autonomous and accepted the emperor as a sort of ideal empirical figurehead. They were subject to the laws of the empire, its jurisdiction, and the resolutions of the Reichstag (parliament). At the same time they participated in the election of the emperor and proceedings of the parliament and could thus influence the politics of the empire through their permanent representatives. Contrary to customs in other countries, the ordinary inhabitants were not subjects of the emperor but of the corresponding territorial sovereigns. After the sixteenth century, these territories included visible geographic outlines.

In the early modern age, the empire could be subdivided into those regions that were strongly bound to it, other zones with a thinned-out presence, and regions on the fringes, the latter being counted as part of the empire, although they did not participate in the political system. Their respective affiliations to the emperor still resulted from the system of medieval fiefdoms and its legal consequences.

These were the boundaries of the Holy Roman Empire: in the north, the duchy of Holstein bordering on Denmark; in the south, the borders with the Austrian countries Styria, Carniola, and Tyrol; in the east, Pomerania and Brandenburg, bordering on the territories of the German Order; and in the west, the borders were contested, especially those with the Netherlands and France.

Charles V was the last powerful emperor of the Holy Roman Empire. He fought against Luther and the Reformation. In his view, Germany was only a sideshow in his Burgundian–Spanish world empire. In the Augsburg peace treaty of 1555, Charles V lost his battle against Protestantism. He had to concede to the German princes their right to decide between Protestantism and Catholicism in their own territory.

From 1583, Emperor Rudolf II resided in Prague, since Bohemia belonged to the Empire.

The Thirty Year War, which lasted from 1618 to 1648, was devastating for Europe, and Kepler would not live to see the end of it. Once again, this war was supposedly a war about religion—in many ways, the last in a long series. No part in the lives of ordinary people and those in power was left untouched by this ongoing affair: economics, personal power, beliefs, employment, and the way to reflect on the world and its destiny. Concerning Europe, the destruction caused by the Thirty Year War was only exceeded several hundred years later by the two World Wars in the first half of the twentieth century—judged by the number of direct and indirect casualties from the sword, famine, and disease.

Initially, the conflict began between Protestant and Catholic states and alliances triggered by the religious policies of the devout Holy Roman Emperor Ferdinand II, but it soon ended up in an already existing France-Habsburg rivalry, involving a number of mercenary armies right across Europe, with Catholic France in the end even joining the Protestant Union in their struggle against the Catholic League. The war erupted full scale after the Protestant King of Sweden, Gustav Adolph, came to the rescue of the faltering Protestants in Germany. The war finally ended with the Westphalia peace agreements in Muenster and Osnabruck, changing the political landscape in Europe entirely in favor of France and Sweden. The results of this war were freedom of religion, destruction and impoverishment of the Germanic countries,

France becoming the most powerful country in Europe, and the Holy Roman Empire being reduced to a meaningless formality.

World View

When Johannes Kepler was born in 1571, people were still influenced by a concept which divided their world into three hierarchical levels. This can best be expressed by the stage representations of public open air theatres (Fig. 3.2).

In these mystery plays, spectators were involved in the drama. The whole stage setting represented the world as perceived by everyone in real life. Reality was divided into the hierarchical levels of heaven, Earth and hell. The theatrical drama and thus life itself evolved between these. And between and among them, an astronomer like Kepler was supposed to interpret what he saw through the magnifying glass of his telescope, in a way that would fit this world without contradiction.

Fig. 3.2 Mystery play in the Late Middle Ages with symbolic scenes of heaven, Earth, and hell

Natural science had slowly set out along its path to deduce its insights from objective observations rather than from the literature of the Ancient Greeks and similar sources, and thus to detach itself from subjectivity. However, the desire for harmony was still dictated by the need to incorporate certain spiritual concepts, those that dominated society and everyday life at the time, into any world model that happened to be conceived. These were the boundary conditions from which Kepler had to extricate himself in the course of his endeavors.

Chapter 4
Childhood and Youth

1571–1594

The order of things, as outlined in Chap. 2, seemed to have been one impetus for Kepler's quest. However, this went hand in hand with his attempt to discover what might be lying beneath it, that is, what was really causing it—in other words, the world behind the world as we perceive it through our limited senses. For Kepler, discovering the inherent order of the world and the secrets behind it were one and the same thing: world harmony. To illustrate the problem, it is useful to consider the following example.

In the Seventh Book of Plato's Politeia, we find his Parable of the Cave [12]. In the first section, an imagined dialogue takes place between Socrates and one of his pupils, Glaucon. The discussion is about the situation of some mysterious prisoners:

Socrates: Imagine this: People live under the earth in a cave like dwelling. Stretching a long way up toward the daylight is its entrance, toward which the entire cave is gathered. The people have been in this dwelling since childhood, shackled by the legs and neck. Thus they stay in the same place so that there is only one thing for them to look at: whatever they encounter in front of their faces. But because they are shackled, they are unable to turn their heads around.

Socrates: Some light, of course, is allowed them, namely from a fire that casts its glow towards them from behind them, being above and at some distance. Between the fire and those who are shackled there runs a walkway at a certain height. Imagine that a low wall has been built the length of the walkway, like the low curtain that puppeteers put up, over which they show their puppets.

Socrates: So now imagine that all along this low wall people are carrying all sorts of things that reach up higher than the wall: statues and other carvings made of stone or wood and many other artifacts that people have made. As you would expect, some are talking to each other and some are silent.

Glaucon: This is an unusual picture that you are presenting here, and these are unusual prisoners.

Socrates: They are very much like us humans, I responded.

© Springer Nature Switzerland AG 2020
W. Osterhage, *Johannes Kepler*, Springer Biographies,
https://doi.org/10.1007/978-3-030-46858-3_4

Socrates: What do you think? From the beginning people like this have never managed, whether on their own or with the help by others, to see anything besides the shadows that are projected on the wall opposite them by the glow of the fire.

Glaucon: How could it be otherwise, since they are forced to keep their heads immobile for their entire lives?

Socrates: And what do they see of the things that are being carried along? Do they not see simply these?

Glaucon: Certainly.

Socrates: Now if they were able to say something about what they saw and to talk it over, do you not think that they would regard that which they saw on the wall as beings?

Glaucon: They would have to.

Socrates: And now what if this prison also had an echo reverberating off the wall in front of them? Whenever one of the people walking behind those in chains would make a sound, do you think the prisoners would imagine that the speaker were anyone other than the shadow passing in front of them?

Glaucon: Nothing else, by Zeus!

Socrates: All in all, I responded, those who were chained would consider nothing besides the shadows of the artifacts as the unhidden.

Glaucon: That would absolutely have to be.

The parable continues with the assumption that some of the prisoners are subsequently set free and hence able to look around and see the fire and the real objects behind them, but still continue to believe in the shadows and disregard the reality behind them. In the end, one of them is released from the cave altogether. Getting into the open, he looks at the Sun, only to get his eyes burnt, but now convinced that he has finally witnessed the real world. When he returns to his fellow prisoners in the cave and tells his tale, they will not believe him, and he is in danger of being killed by them.

This was not quite what would happen to Johannes Kepler, but he certainly made up his mind, even as a young man and throughout his whole life, to discover the world behind the world, to discover the secret workings of the world, of which most people seem only to see reflections, just as those prisoners only saw shadows.

Weil der Stadt

Johannes Kepler was born in a small town, a Catholic enclave in Protestant Baden-Wurttemberg called Weil der Stadt (the Town of Weil, Fig. 4.1), the second smallest free imperial city in the Holy Roman Empire, on December 27th 1571 according to the Julian Calendar (the Gregorian Calendar was only introduced in 1582). He was baptized a Catholic, but educated later as a Protestant.

Fig. 4.1 Weil der Stadt marketplace and the Kepler monument

Weil der Stadt is a small municipality in Baden-Wurttemberg in Böblingen County in the administrative region of Stuttgart. The region is a famous wine-producing and fruit-growing area, dominated by beautiful rolling hills. All the towns mentioned below, where the young Kepler spent his childhood, are in the same area, except for Tubingen, which is situated some 50 km further south. Today this region is home to the mighty car industry of Daimler-Benz.

The house in which Kepler saw light for the first time still stands near the market place, although it is a replica, built exactly like the original, shortly after the latter was destroyed during the Thirty Year War. In this same market place, a prominent monument was erected in his honor in 1870. Kepler's was a premature birth, when his mother had been carrying him for only seven months, so he was a weak child from the very beginning.

In the family chronicle which he wrote at the age of twenty-five while working as a teacher in Graz, he described his early encounter with astronomy. He remembered that his mother took him to a hilltop to observe a comet in 1577, at the age of six. And when he was nine years old, he observed a lunar eclipse.

The house in which he was born belonged to his grandfather Sebald Kepler, who was lord mayor of the town, with its population of just two hundred souls. According to his own account, his grandfather was temperamental and authoritarian, his grandmother disruptive and annoying. She had given birth to twelve children, of whom nine survived, and among them Kepler's father Heinrich. In the chronicle, Kepler drew a devastating picture of his parents as well: his father was something of an adventurer and had enlisted with the Catholic Duke Alba to fight the Calvinists, despite his Protestant denomination. His mother was known all over town as a cantankerous individual.

He spent his childhood in the narrow confines of his grandfather's house, together with his siblings and the families of his aunts and uncles. His bodily weakness was aggravated by a series of illnesses and diseases. When he was four years old, he caught smallpox. Although he survived, he continued to feel the consequences of this disease for the rest of his life, suffering from headaches and attacks of fever. Stomach and gallstone problems forced him to adhere to specific diets, and hemorrhoids prevented him from remaining seated for any length of time. To top it all, he suffered from polyopia, a form of multiple vision, which was particularly awkward for his later astronomical observations.

Leonberg and Ellmendingen

In 1576, his parents relocated to Leonberg, since his father's adventures in the Netherlands had made it impossible for him to stay in Weil der Stadt. They remained there until 1579, but Kepler's father left the family again in 1577 for new adventures in Belgium. When he returned in 1579, he was broke and they had to sell the house. The next stop was Ellmendingen near Pforzheim, where he leased an inn called "Zur Sonne" ("Towards the Sun") in 1580.

Since Johannes would never have been able to exercise a profession requiring bodily strength, his mother decided to enlist him in the local school in Leonberg in 1577. This provided courses for reading and writing in German with an option for talented pupils to join a Latin stream later. Kepler was successful. After three classes in the Latin stream, pupils had to pass the so called "Landesexamen" (Country Exam) in Stuttgart. However, Kepler's studies were interrupted because of the move from Leonberg to Ellmendingen in 1579, when he had just been promoted to the second class. The boy succeeded in catching up and passed the Landesexamen successfully on May 17th 1583, while not quite twelve years old.

By then, his father had also been successful in his inn business and decided to move back to Leonberg to buy a new house there. From there, he sallied forth to serve the Neapolitans and was never seen again thereafter.

Further Education

Although the entrance age was normally fourteen, Johannes was already admitted in 1584 at the age of thirteen, on recommendation by his teachers in Leonberg, for further education in the former Premonstratensian monastery in Adelberg. Two years later he was admitted to the former Cistercian monastery in Maulbronn to continue his studies, and he stayed there for three years. It was during these years of study that he first developed an awareness of the arguments for and against the three major confessions Catholicism, Lutheranism, and Calvinism. He was particularly worried about the doctrine of predestination (an in-depth assessment of his internal and external conflicts arising from these considerations will be presented in Chap. 7).

At the age of seventeen, he was capable of passing the baccalaureate exam in Tubingen. He had to return to Maulbronn for another year to finish his studies as a so-called veteran. Only in autumn 1589 was he finally admitted to the Tubingen seminary. There, the candidates had to attend lectures in the arts for two more years before passing their master's exam as another precondition for further theological studies over the next three years. Since this whole education was supported by stipends, successful theologians had to remain in the services of the duchy for the rest of their lives. Only with special permission from the prince could someone practise outside its borders.

At the age of nineteen, Kepler passed his master's exam. During his studies he had also been busy acquainting himself with the writings of Platonic and neo-Platonic scholars influenced by the school of Pythagoras.

Pythagoras

None of the original written works of Pythagoras have been preserved and no copies of them have been transmitted down through the centuries since he lived and worked. The most important biographers of this enigmatic person—Diogenes Laertos and Porphyrios—lived more than seven centuries later. For this reason, legends about him and his role in philosophy, science, and society abound. Some people thought of him as a great mathematician, others were of the opinion that he did not understand anything at all about mathematics and was just the leader of an obscure sect. Aristotle claimed that Pythagoras was a charlatan, and Heraclitus that he was a cheat. What is known is that he entertained a close circle of followers committed to high moral standards, vegetarian alimentation, and loyal friendship within the group.

Pythagoras was born around 570 B.C. on the isle of Samos. His family were traders. Samos was situated near the coastal town of Milet, and it seems likely that Pythagoras met Thales, who lived there at that time. Even as a young person he travelled a lot, including a visit to Egypt in 547. Some historians believed that Pythagoras was initiated into astronomy and geometry during a twenty year stay in Egypt. He returned to Samos in 513 and founded his famous school there. Besides their interest in mathematics, they occupied themselves with the theory of music, anamnesis of prior existences, and metempsychosis. The group relocated to Kroton when Pythagoras was forty years old. Later, due to war, they had to flee from there to Metapont, where he died in 497.

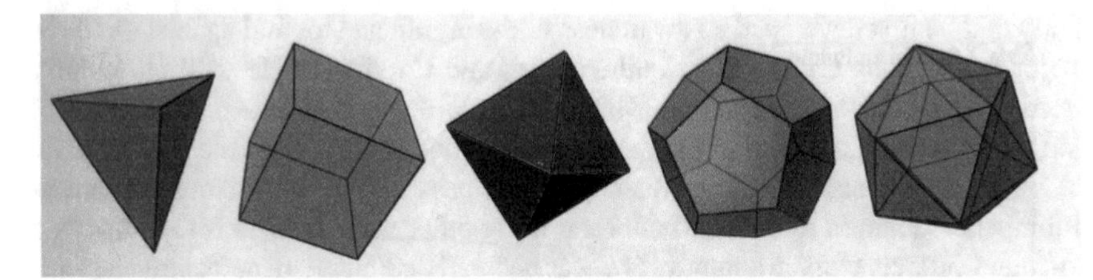

Fig.: 4.2 The five Platonic bodies

The teachings of the Pythagoreans continued to influence people throughout the whole of the Mediterranean. Even Ciceró reported that he had visited the house (in 78 B.C.) in which Pythagoras had lived and died.

The following findings are attributed to the Pythagoreans:

- A philosophy of numbers and number theory (even, uneven, prime, and coprime numbers)
- The Pythagorean theorem
- Angle rules for parallels
- The principle of surface transformations
- The theory of proportions
- The regular polyhedrons
- A variety of arithmetic formulas
- Means (arithmetic, geometric, harmonic)
- Polygon theory.

There were two subjects which particularly drew Johannes Kepler's attention when he studied the Pythagoreans: the regular polyhedrons and their theory of music. The three regular polyhedrons known to them were the cube, the tetrahedron, and the dodecahedron, and these were later supplemented by Theaitetos, who added the octahedron and the icosahedron. All five were later known as the Platonic bodies (Fig. 4.2). They became a major pillar in Kepler's harmony of the world.

The Pythagoreans had discovered that musical intervals could be represented by simple ratios of numbers. They applied their interval theory to planetary orbits to postulate a celestial harmony, which was later referred to as the music of the spheres by Eudoxos. Plato took this theory over in his Politeia.

Maestlin

During his studies in Tubingen, Kepler developed a special relationship with one of his professors: Michael Maestlin, an astronomer and mathematician—a relationship, which would last for many years, even after he had finished his studies. In fact, Maestlin was instrumental in Kepler's later theological disputes with the Wurttemberg clerical establishment, which will be dealt with in detail in Chap. 7.

Michael Maestlin was born in Goppingen in 1550 and died in Tubingen in October 1631. After studying Protestant theology, mathematics, and astronomy in Tubingen itself, he became deacon in Backnang, and thereafter professor of mathematics in Heidelberg. Then from 1583, he taught mathematics in Tubingen. He was a strong supporter of Copernicus' heliocentric world model, and it was Maestlin who first acquainted Johannes Kepler with Copernicus' opus major "De revolutionibus orbium celestium" and with Euclidean geometry. However, Maestlin's support for Copernicus initially remained confined to his inner circle, since his theological colleagues frowned upon it. Kepler himself decided not to hold back and entered into public debate over it.

Nicolas Cusanus

Besides the Ancient Greeks, Pythagoras and Euclid, and Copernicus, Kepler drew from another important source, a universally acclaimed genius of his time: Nicolas Cusanus or Nikolaus von Kues, in the German version of his name. Kues is a romantic town in the lovely Moselle valley, today called Bernkastel-Kues. The famous philosopher, theologian, and mathematician was born there in 1401. He belonged to the early German humanist movement and later became cardinal and vicar general in the Papal States.

In Cusanus' scientific theory, he tried to find analogies between mathematical and mystical thinking, emphasized in his work "De mathematica perfection" ("On Mathematical Perfection"). His mathematical magnum opus was entitled "De mathematicis complementis" ("On Mathematical Complements").

More important and in essence far more revolutionary than Copernicus were his cosmological conclusions. Far ahead of his time for centuries he claimed that:

- The Universe has no limits, since its boundaries could not be observed.
- The Earth is not situated at the centre of the world, and it is not at rest but in motion. It is just another star, without being a perfect sphere.
- The other celestial bodies do not orbit in circular fashion but rather along ellipses.

Cusanus was also the first person to postulate a multiverse, containing a multitude of different worlds, but integrated into a super-universe.

However, although Cusanus broke with the geocentric world model of his time, he was not a predecessor of Copernicus, since he did not put the Sun at the centre of the world either. He went even further. For him, there was no centre of the world at all, nor absolute motion, since there existed no stationary reference system. He thus came very close to Einstein's general theory of relativity, five hundred years ahead of his time!

Johannes Kepler was so impressed by Cusanus' geometrical mysticism that he referred to him as "divus" (divine) in his debut publication "Mysterium Cosmographicum" ("The Cosmographic Mystery").

Leaving Tubingen

Kepler was supposed to complete his theological studies in 1594. However, before this happened, a request arrived from Styria, asking the senate of Tubingen University to propose a mathematician for a teaching post at the Protestant convent school in Graz. The candidate was Kepler. This meant that he could forget about his career as a preacher. After some soul-searching he accepted the offer. At the age of twenty-two he left Tubingen on March 13th 1594 in a carriage headed toward the distant Graz, where he arrived on April 11th. He would only see Tubingen again on sporadic visits in his later life. It is not quite clear, why the senate decided to choose Kepler to satisfy the request from Styria, since the young man was close to finishing his studies. There may have been two underlying reasons, apart from the question of competence: Kepler's leaning towards Copernicus and the suspicion that he might have sympathized with the Calvinists. This and Kepler's inclination to position himself above the three confessions Catholicism, Lutheranism, and Calvinism in a rigid and unfaltering attitude may have influenced Maestlin and the senate in their decision. So Kepler profited from the so called "beneficium emigrationis", a provision granted only exceptionally to people who did not conform to the confession of the Prince and wished to emigrate to a more convivial place.

Chapter 5
Graz

1594–1600

When Kepler consolidated his "Mysterium Cosmographicum", he amalgamated three threads of human reasoning: mystical, religious, and scientific. In the end this composition led him to his "Gesamtkunstwerk" (Total work of art). It is intriguing to find out that in one way or another this line of reasoning continues still in our own time.

In his work "Einstein, Gertrude Stein, Wittgenstein & Frankenstein" [13], John Brockmann writes that in quantum theory Heisenberg's uncertainty principle makes it impossible to determine complementary properties of elementary particles, such as position and momentum. This is of course scientifically true. He then continues that it does not make sense to exclude the observer from the observed. This is again true. And finally he muses on non-locality, i.e., properties of one part of a quantum system influencing those of another part, when measured, in such a way that they seem to communicate with each other without time delay.

Brockmann then cites the physicist and philosopher David Bohm, who tried to bring some sense into the debate. According to Bohm, the reason we cannot comprehend phenomena treated by physics belonging to things and events from our own world of experience is that we do not acknowledge that they are part of some sort of totality or wholeness determined by a higher dimensional sphere of implicit order. What is déjà vu about this is that Kepler thought along exactly the same lines—without of course having any notion of quantum physics.

The mathematician G. Spencer-Brown went even further. He claimed that we can now construct the Universe as we know it today, in every detail we could wish for. However, whatever we construct will never be the whole, since while constructing it, the Universe will continue to undergo changes which we will only be able to integrate into our picture at a later date, and so on and so forth. So in a sense, we shall never be able to catch up.

Concerning the role of religion in Kepler's quest, the physicist Paul Davies raised the following questions in 1986:

© Springer Nature Switzerland AG 2020
W. Osterhage, *Johannes Kepler*, Springer Biographies,
https://doi.org/10.1007/978-3-030-46858-3_5

Why are the laws of nature as they are?
Why does the Universe consist of just this raw material?
How did this raw material come into existence?
In what way did the Universe obtain its order?

Bohm doubts that modern science will be able to answer these questions decisively, and cites the example of the incompatibility of quantum physics and relativity. He believes that, in the way it is practised today, physics is so fragmented that it is further away than ever from an answer regarding any kind of wholeness of the world, and thus also from Kepler's achievements.

Carl Friedrich von Weizsaecker took a quite different approach. He is famous for inventing the liquid drop model of the atomic nucleus and a formula that could be used to calculate the binding energy of its constituents. The latter thus allowed one to calculate the energy released during nuclear fission, which eventually contributed to the development of the atomic bomb. In 1948, von Weizsaecker published a book entitled "Die Geschichte der Natur" (The History of Nature) [14], which contained twelve lectures given at the University of Gottingen. He attempted to describe the wholeness of nature and the world, including the union of the humanities with the natural sciences. His starting point was the risk that specialization of the sciences would lead to a fragmentation, with the result that science would no longer be able to develop a unified world model. In his view, modern scientists were no longer able to know everything from all branches. They had to relinquish that claim and thus specialization was the result. This was therefore the fate of science as such.

On the other hand, a scientist is not only a scientist but a human being as well, with convictions, both moral and religious. His instrumental knowledge is made up of fragments due to the separation of subject and object in his thinking, a separation man must have achieved at some time during his evolution, when he became aware that the outside world and his own consciousness were two separate things, although there exists an affinity between the two things. In the end this separation is what led to the parallel development of the humanities and the natural sciences. Natural science uses instrumental approaches to describe the material world, whereas the humanities try to do the same by exploring man: his consciousness, spirit, and soul. Von Weizsaecker states that nature came before man, but that man himself came before the natural sciences. On this basis, he then tries to reconcile the humanities with the natural sciences. He takes the reader on a voyage through the history of the Earth, the structure of the cosmos in space and time, its infinitude, the formation of the stars, and finally life itself. His approach was thus once again a contructivist one, just as Kepler's had been in his day, when at least science seemed to be a whole—in contrast to today's rather deconstructivist tendencies.

Fig. 5.1 Graz in ancient times

Graz

Graz is today the second largest town in Austria and provincial capital of Styria. It is situated in the Graz basin on the river Mur. In 1379 it became a residence of the Habsburgs (Fig. 5.1). Apart from Styria, they reigned over Carinthia and parts of Italy, including Trieste, and parts of Slovenia including Carniola. After 1520, the population became Protestant under their mayor Simon Arbeiter, a chemist. Between 1573 and 1600, two institutions competed for the education of young people in Graz. The first was the Catholic University with its associated grammar school, the Archiducale Gymnasium Soc. Jesu Graecensis, which became Academia, Gymnasium et Universitas. The other was the Protestant monastery school, formerly the Santa Clara Monastery School. It was the latter which invited Johannes Kepler to join as a teacher. He succeeded the mathematician Georg Stadius. His annual salary as professor was small, amounting to 150 Guilders.

Prognostica

Kepler was supposed to teach logic, metaphysics, ethics, mathematics, and astronomy. He gave up mathematics, since attendance was poor in his first year and nil in the second. Instead, he had to teach Virgil and rhetoric.

However, Johannes Kepler initially became famous, not for any scientific achievement in the sense we would use the term today, but through the publication of his "Prognostica" (Predictions). This appeared in a yearly calendar based on the rules of astrology, which Kepler had made himself familiar with. This brought him an additional income of about 20 Guilders. He published this calendar for six consecutive years. Kepler himself was not quite convinced about their scientific value and adopted a cautious attitude toward them. In a later statement to the personal physician, Feselius, of the Margrave of Baden he called astrology "the silly baby girl of the wise mother astronomy".

Modeling the World

In 1595, Kepler became acquainted with the 23 year old Barbara Mueller, daughter of the well-off Jobst Mueller from Muehleck zu Goessendorf, south of Graz. She had already been married twice despite her young age, but both her previous husbands had died after only a few years of marriage. Before Kepler finalized their relationship, however, he returned to Wurttemberg for seven months to work on his "Mysterium Cosmographicum".

Kepler had been a fervent admirer of the Copernican world model ever since he had been introduced to it by his mentor Maestlin in Tubingen. At the same time he positioned himself against the traditional model upheld by Ptolemy. While contemplating Copernicus, he was concerned with the question of a deeper explanation for the distances between the planets. The problem was somewhat simpler than the same question today, since at the time there were only six planets going round the Sun in the heliocentric configuration in question: Mercury, Venus, Earth, Mars, Jupiter, and Saturn. However, it was complicated enough even then.

The Quantization of the Solar System

Thus Kepler concluded beforehand that there should be physical laws governing the orbits of the planets, laws that would make them go round the Sun on precisely that path they were seen to be travelling on and no other—preferred and predetermined trajectories. This sounds highly analogous to the semi-classical model of the atom successfully proposed more than 320 years later by Niels Bohr. His explanation for the motion of electrons around the atomic nucleus was that there existed allowed orbits for their motions on which they would not lose their kinetic energy due to the constant dipole radiation predicted in normal classical systems. This finally led to the concept of quantum numbers, and bringing in Planck's constant, the electron orbits could be calculated.

Looking back to Kepler's time, should it not have been possible to find a similar mathematical mechanism to arrive at allowed orbits for the planets in our Solar System–leaving Planck and quantum theory aside? The answer is: not really. If we take Bohr's equations for the radius of the hydrogen atom and replace the electromagnetic potential by gravitation, we find no equivalent for Planck's constant and thus no universal yardstick, but different constants, one for each known planetary orbit.

Applying de Broglie's matter wave theory for particles to planets leads to wave lengths far below the Planck length (1.6616×10^{-35} m), which means that particles or bodies above a certain size can only be observed as matter, but never as waves.

Thus, having failed by two approaches to deconstruct our Solar System using modern methods, about which Kepler knew nothing, was there really no way at all for him to arrive at his declared aim of explaining the structure of the world by mathematical means?

Johann Daniel Titius and Johann Elert Bode

Johannes Kepler tried initially to find the answer to his planetary distance problem by numerical combinations. For example, he tried to find out whether one or the other "spherical" distance was by any chance a multiple of another one—without success. It was about 170 years later, in 1766, that a numerical solution was proposed by Johann Titius, a German astronomer and physicist, published by Johann Bode in 1772. Titius had derived an empirical formula which could be used to calculate the distance between a planet and the Sun from its sequence number when moving away from the Sun. The formula itself is simple mathematics: the mean radius of a planet's orbit around the Sun is

$$R_n = 4 + 3 \cdot 2^n$$

where n is the sequence number of the planet in question, counting from Mercury and moving outward. The result is not the absolute value, but a value relative to the Earth's.

Within an error of a few percentage points this formula gives the correct values—with the exception of Mercury and Neptune. Since no physical explanation for the validity of this formula could be found, it was later decried as simple number gimmicks. However, it has been revived in recent times to analyze extra-solar planetary systems.

Euclid

Johannes Kepler had studied Euclidean geometry, and this became yet another source that would influence his thinking and theorization.

It is reported of Johann Wolfgang von Goethe that he explained Euclidean geometry to the pedagogue Johannes Daniel Falk in the following way:

> Geometry here is conceived from its first elements, just as Euclid presents them to us, and we start off just like beginners. Thereafter, however, she becomes the perfect preparation for, even introduction to philosophy. [15]

Euclid was born in Alexandria in about 360 B.C. He was educated in Athens and later called back to Alexandria by Ptolemy I Soter. The authorship of the works attributed to him is disputed. Some scholars claim that these are really a collection of writings later edited by his pupils under his name—a procedure not uncommon in antiquity and also used for biblical books (prophet Isaiah, some epistles from St. Paul, etc.).

Euclidean geometry is laid out in 15 books, the "Books of the Elements". Further works attributed to him include "Data", "About Segmentation", "Optika", "Katoptika" (optical illusions), "Sectio Canonis" (theory of music) and "Phainomena" (astronomy). Euclid's single most important contribution to mathematics was his attempt to base geometry on axioms and thus to introduce first principles into mathematics. One of the most famous is the parallel postulate, which reads:

> If a line segment intersects two straight lines forming two interior angles on the same side that sum to less than two right angles, then the two lines, if extended indefinitely, meet on that side on which the angles sum to less than two right angles.

This axiom cannot be deduced or proven from the other Euclidean axioms. As a consequence, mathematicians like Janos Bolyal, Nikolai Lobachevsky, and Carl Friedrich Gauss tried to develop a non-Euclidean geometry. This was based on two possible alterations of the original parallel postulate.

Either: "There exists no parallel to a straight line passing through a point outside this straight line."

Or: "For a straight line and a point outside the straight line there exist at least two parallels."

It was Gauss's pupil Bernhard Riemann, who later developed non-Euclidean geometry further, and this led to his differential geometry of curved space. The latter in turn served as a toolkit for Albert Einstein to develop the general theory of relativity, the basis for modern cosmology.

This brings us back to Johannes Kepler and his attempts to build a cosmological model. He also based his considerations on Euclidean thinking, albeit from a somewhat different point of view. It had to do with the regular polyhedrons, already mentioned in the preceding chapter, when we discussed Pythagoras. Initially, ancient Greek philosophy allowed only for the existence of four regular polyhedrons corresponding to the four supposed elements: fire, earth, air, and water. Euclid dealt with these geometrical bodies in his 8th and 11th Books of the Elements. He dwelt particularly on the question of how to construct them from regular congruent polygons.

Kepler had given up his attempt to construct a world harmony from pure numbers, of the kind that Titius would imagine later. He had become an expert on Euclidean geometry and tried to find a solution to his problem via this route. He was convinced that God had created an archetype of geometry even before the creation of the world. Euclid must have sensed that, since he concluded his work with a description of the wonders of the five regular polygons. Kepler was convinced that these must be connected in some way with the six planetary spheres.

Mysterium Cosmographicum

On July 9th 1595, a sudden insight into the relationship between the six planetary spheres and the five regular polyhedrons came during a lecture he was giving. Ten days later, on July 19th, he already pronounced his first results. If polyhedrons possessed an outer sphere circumscribing them and at the same time an inner sphere inscribing them, the task at hand was to find a solution to the following problem: could the regular polyhedrons be inserted between the planets in such a way that the sphere on which one planet travels circumscribes one polyhedron and inscribes another polyhedron on which the next planet travels, and so on?

To find his solution he had to deal with a hundred and twenty possible permutations or arrangements. The best approximation for the relation between the planetary orbits resulted in the following sequence (Fig. 5.2):

- The Earth as internal sphere of the dodecahedron
- Mars as external sphere of the dodecahedron, which serves at the same time as the internal sphere of the tetrahedron
- Jupiter on the external sphere of the tetrahedron, which serves as internal sphere of the cube
- Saturn on the external sphere of the cube
- The Earth at the same time on the external sphere of the icosahedron
- Venus on the internal sphere of the icosahedron, which serves at the same time as the external sphere of the octahedron
- Mercury on the internal sphere of the octahedron.

And there exists a further relationship between geometrical bodies and the five elements of the Ancient Greeks:

- Fire for the tetrahedron
- Earth for the cube
- Air for the octahedron
- Water for the icosahedron
- Ether for the dodecahedron.

When he had finished this tedious work, Kepler believed that he had lifted the veil on God's glory, which had hitherto been shrouded in human ignorance. He had run through all possible permutations, although for some of them their unsuitability had been obvious at the outset. The long months of the winter of 1595–96 passed in this way. Kepler was exhausted and set off for a period of convalescence in Wurttemberg, with his two grandfathers, who were still alive. His most important call was on his earlier mentor and teacher Maestlin, to seek his advice on what he had achieved. Maestlin's judgment was unequivocal: he declared Kepler's world

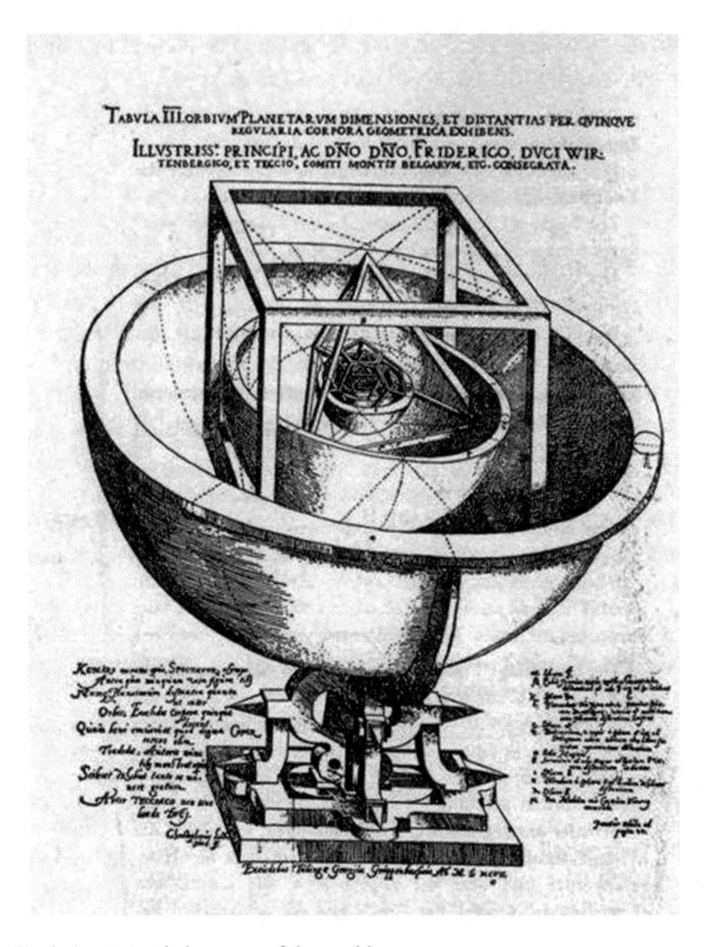

Fig. 5.2 Kepler's geometric harmony of the world

model to be adequate and in perfect order. Not a single element of it could be altered without risking a complete collapse of the whole system. Another of his previous professors was Matthias Hafenreffer, with whom he later entered into a major theological dispute (Chap. 7). Hafenreffer advised him to refrain from trying to prove the consistency between Copernicus and the Holy Scriptures in his own book, but rather to concentrate on purely mathematical facts.

The title of his work was long—in accordance with the custom of his times. It read:

> Prodomus dissertationum cosmographicorum continens Mysterium cosmographicum de admirabili proportione orbium coelestium; deque causis coelorum numeri, magnitudinis, motuumque periodicorum genuinis et propriis, demonstratum per quinque regularia corpora geometrica.

(Predecessor of cosmographical tracts comprising the secret of the world regarding the wonderful relationship between celestial objects, and about the hereditary and proper causes regarding the number and the periodic movements of celestial objects, proved by the five regular geometric bodies).

It did indeed turn out to be a "predecessor", as later developments would prove. It concluded with a final hymn to God, the Creator. This hymn sounds just like a psalm, praising all living creatures, including man. In the middle of it, Kepler referred to his own endeavors:

> ....
>
> But I look for the traces of Thy spirit out in the Universe,
>
> I regard ecstatically the glory of the mighty celestial edifice,
>
> This artful work, the mighty wonder of Thy almightiness.
>
> I see how Thou hath determined the orbits according to fivefold norms,
>
> With the Sun in the middle to donate life and light,
>
> I see that her laws regulate the course of the stars,
>
> How the Moon achieves her transitions, suffers eclipses,
>
> How Thou scatterest millions of stars across the realm of the skies.
>
> ....

Before publication, the agreement of the senate of the University of Tubingen had to be secured. Maestlin took care of that and arranged for it to be printed during the winter of 1597. In spring 1597 when he returned to Graz to marry Barbara Mueller, Kepler held the first copies in his hands.

Galileo Galilei [16]

Kepler sent one copy of the Mysterium to Galileo Galilei, who was then 33 years old. Galileo was professor of mathematics at the University of Padua. Although the Italian had not yet published anything of significance, his lectures and inventions had brought him renown far beyond the borders of Italy. Galileo was born before Kepler and died after him, so Galileo had been influenced by the same historical upheavals as Kepler, although some of the events had had different impacts, since the situation in Italy was not the same as in the ongoing disintegration of central regions of Europe.

For Galileo, there were inventions and discoveries, there were attitudes and teachings, and there was posthumous glory acquired by controversy and suffering at the hands of the powers of his time. Leaving aside the latter details, his tangible achievements have to be weighed up against the technological and scientific resources available during his lifetime.

Galileo has been credited with the following major inventions:

- Hydrostatic balance
- Galileo's pump

- Pendulum clock
- The sector
- Galileo's thermometer
- Telescope.

His inventions not only demonstrated his genius as a scientist, but were also proof of his engineering skills. Although he engaged craftsmen to construct his devices, he certainly did lay hands on himself when designing his instruments. In any case, he devised and supervised the technology in question. This was also true for improvements he made to certain simple laboratory devices like the inclined plane. In some cases, he undertook engineering work as an extra source of income. On the other hand, although he was credited with certain inventions, they were not always entirely of his own making.

The hydrostatic balance was inspired by the tale of Archimedes' discovery of the specific weight. And his pump, for which he was granted a Venetian patent, was based on the Archimedes screw. The invention of the pendulum clock is generally credited to Christiaan Huygens, who published a paper on its workings in 1657, after Galileo's death. However, Viviani, Galileo's first biographer, reported that his master had already been working on such a device in 1641, but had been unable to complete these studies because of his blindness. One unique invention is his sector or compass. And his development of a thermometer before the existence of absolute or relative temperature scales can be regarded as a major breakthrough, although his apparatus would better be called a thermoscope.

The history of the telescope is well known and will be discussed in more detail in Chap. 6. Galileo was just someone who improved on the earlier devices, but who certainly takes all the credit for their application to astronomy and the resulting discoveries of new celestial bodies. This takes us from engineering to science itself.

These are his main publications:

- Siderius Nuncius
- Saggiatore
- Dialogo
- Discorsi.

There is no doubt that these were the basis for his fame during his lifetime, long before his conflict with the ecclesiastical authorities triggered the history of his glorification. Galileo was accepted as an eminent scientist in many fields, ranging from philosophy to physics, astronomy, and others. Scientists in those days were of the universal kind, capable of world views encompassing nearly everything, something that in our own times of specialisation would be attributed only to a lateral thinker. However, he was not alone in possessing this qualification. Many of his contemporaries were similarly endowed. On the other hand, there were those like Markus Welser, Lord Mayer of Augsburg, a friend and benefactor of Galileo, who in a letter dated May 20th 1611 begged "to be spared for a while to follow Galileo along the path of the earth's motion, because he had difficulties in shackling his mind in as much."

The really outstanding tangible achievements were his astronomical discoveries, starting with the discovery of Jupiter's natural satellites and his observations of the surface of the Moon. If something like a Nobel Prize in Physics had existed at that time, Galileo would surely have been a candidate for it, for this alone.

The discussion about Aristotle and motion, as well as the competition between the two world systems, is another matter. Here, Galileo's position was not unique. Many other scholars of his time were heavily involved in these quarrels; none of them had proposed the heliocentric system, including Galileo. They could take it or leave it. But the time was ripe. Unfortunately, because of his already well established standing in the scientific world and in society, Galileo's word had quite a different weight, compared with those in inferior positions. On the other hand, his standing in society and connections to the highest ecclesiastical circles not only forced the latter to act as they did, but also in the end protected him from earning a similar fate to Giordano Bruno.

However, there are other, more fundamental aspects, regarding Galileo's contribution to modern science. He also tried to popularise science by using everyday language, the "Volgare", instead of the baroque style of presentation favoured by his contemporaries.

It is unfortunate that the drive for a science promoting the full application of reason, as displayed in all his works, was superimposed by a never-ending discussion of morality, which today seems to have been the only source and foundation of his fame. The discrepancy between Galileo's approach and what was expected of a scholar in those days becomes obvious from one comment in one of the expert opinions drafted for his tribunal, which reads: "The author claims to have discussed a mathematical hypothesis, but at the same time he attributes to it a physical aspect, something no true mathematician would ever do."

Of course, a lot of his thinking and a lot of the struggles he undertook throughout his career centred on Aristotle and his followers. There seems to be no doubt that Galileo saw himself as a pillar of natural philosophy on a par with Aristotle. He was on the same level as the ancient master. Initially, he followed his Florentine teacher Ostilio Ricci, one of a group of scholars who rejected the literalism prevailing in scholastic circles. The critics included people like Johannes Buridanus, a member of Ockham's nominalist circle, Nicolas d'Oresme, Bishop of Lisieux, and Albert of Saxony, a pupil of Buridanus. They were called high scholastics. It was their impetus theory which first attracted the attention of Galileo before he set out to develop his own experimental mathematical method to study the measurable properties of motion.

A further source of stimulation was Giovanni Battista Benedetti, a successor of the Buridanus movement, who had published a tract called "Demonstration of properties of local motion, contradicting Aristotle and all philosophers", which was critical of Aristotle's theory of motion. Benedetti had been a pupil of Niccolo Tartaglia, the mathematics teacher from Brescia who had been interested in free fall and the trajectories of projectiles. All this meant that the young Galileo could refer to both the writings of the ancients and those of the empirically oriented authors of his day.

In later years, his contributions were not limited to opening up regions hitherto regarded as being unchangeable and thus unexplorable from either a mathematical or a physical point of view. He went further in demanding that people should participate in his observations and reports. His sharp eye and deductive capabilities, trained by experiments with inclined planes and swinging pendulums, simple and complicated measuring tools, also made it possible for him to interpret observable changes in the skies. The result was the "Siderius Nuncius".

Even this work found its adversaries, for example in Stelluti, a close collaborator of Cesi, who claimed in a letter that Kepler had written against the Siderius. This was far from true, since Kepler had given his assent to Galileo in his "Dissertatio cum Sidereo" in March 1610. The "Dissertatio" was a sort of public letter, in which Kepler expressed his support for Galileo. Kepler had received the Siderius on April 8th and issued his Dissertatio only 22 days later. Kepler went even further, speculating that the Moon and Jupiter might be inhabited and suggesting that man would one day be able to travel to these celestial bodies and visit them. He also wondered about the possible influences the new celestial bodies would have on astrology. Galileo ignored these speculations, but nevertheless expressed his gratitude for Kepler's support in a letter, stating that he thanked him "because you were the first, and practically the only one, to have complete faith in my assertions." Kepler himself was unable to verify Galileo's observations himself, because he lacked the appropriate instrument, and Galileo had ignored his request for a better telescope. In the end, a year later, Kepler had to borrow one from Giuliano de Medici, a Tuscan resident in Prague who had received the telescope from Galileo as a gift.

Another bone of contention between Galileo and Kepler concerned the elliptical path of the planets proposed by Kepler in his Astronomia Nova. In communications with Cesi, the founder of the Accademia dei Lincei, Galileo rejected Kepler's solution.

Galileo was convinced that his methods would allow for impartial exploration of nature, as he once put it in a letter to Cesi. When it came to the crunch later on, he tried to defend his position against the assaults of ecclesiastical scholars by proposing a clear separation between the scope of the sublime (a phrase from Nikolaus von Kues), i.e., the divine and revelation, and the scope of natural phenomena, where scientists should be left to their own devices. Neither side should interfere in the scope of the other. In a communication to the Archduke's mother Cristina di Lorena in 1615, he explained that theology can indeed claim to be of "highest authority", but only concerning its proper subject. It therefore should not try to deduce from this the claim that it is in the possession of absolute truth. Further on, he asserted:

> Therefore, its professors should not arrogate the authority to give orders in professions which they have neither practised nor studied.

He compared the role of theology to that of an absolutist prince, who should have no interest in dealing with the day-to-day business of medics and builders, since he would endanger people or buildings through his lack of specialised knowledge.

His framework for his thinking was based on effective rational grounds. In the "Dialogo", he proclaimed the equality of Earthly and heavenly motions, deducing that the motion of the Earth was a completely normal occurrence—even before describing it mathematically. At the end, he achieved the transition from speculative to verifiable science. The abstract concept of matter was replaced by the notion of its experimental objectivity as a system of physical forces. This approach and its presentation as a method of research in natural science remains a common theme throughout his work and has since become accepted as generally valid.

Against Sarsi, one of his rivals, he wrote:

> Philosophy is written in this grand book—I mean the Universe—which stands continually open to our gaze, but it cannot be understood unless one first learns to comprehend the language and interpret the characters in which it is written. It is written in the language of mathematics, and its characters are triangles, circles, and other geometrical figures, without which it is humanly impossible to understand a single word of it; without these, one is wandering around in a dark labyrinth.

This was written as early as the Saggiatore, and some people claim that the Saggiatore was the first manifesto of natural science.

Although Galileo's tangible scientific results were never forgotten, the mention of his name invariably invokes one thing: his conflict with the church and thus the evil the church did to science and mankind as a whole, in a way that discredited the church as a reasonable or rational institution.

During the nineteenth century, a number of biographies and essays about Galileo came out, each trying to categorise him as a symbol of either the Enlightenment or anticlericalism under morally positive or negative appraisals. One famous theatre play is Brecht's "Galileo Galilei", depicting its chief character as being scientifically obsessed and later manipulable by those in power. But Brecht's drama was no more historically correct than Schiller's "Wallenstein's Camp".

This logic was used throughout the development of the Enlightenment and is a standard reflex today. However, there are other voices trying to provide a somewhat different interpretation. Galileo was rehabilitated on November 2nd 1992 by Pope John Paul II. The records of his trial were made accessible to science in 2008 and some interesting conclusions can be drawn.

First of all, it seems that both were in error: Galileo on the scientific side and the Curia on the question of theology. The Inquisition did not realise that the contradiction between heliocentricity and the scriptures was only apparent. And Galileo was in no position to prove scientifically that his or the Copernican world model was in fact correct. It took a long time for the Inquisition to act and demand Galileo's revocation of his conviction that the Sun was at the centre of the Universe. But the core of the argument was not obvious. Anyone who cares to look can see that the Sun does indeed rise in the east and set in the west every day. It does not stand still. The other obvious thing was that an Earth moving around was simply not part of everyday experience either.

In 1561, after Copernicus‘ publication, both world models, neither of which were proven scientifically according to today’s standards, were taught in Salamanca, Spain, for example. The Inquisition initially tried to nudge Galileo into being more careful with his apodictic statements. But in the end Galileo was too proud to adhere to this advice and it thus came to an official trial. And incidentally, the verdict was never signed officially.

In fact, the Copernican world model had never been a serious problem for the church and the Pope. Galileo tried to prove the motion of the Earth by resorting to the phenomenon of the tides. But the tides were a bad example, unable to deliver the truth. This is standard knowledge today, and on this point the Inquisition was therefore right. Galileo even rejected the proposal that his results be categorised as a simple hypothesis. This suggestion by Bellarmin is nowadays common practice: every scientific assertion should initially be regarded as a hypothesis. This notion was first formulated by Karl Popper in 1934 in his “Logic of Scientific Discovery”. Galileo did not accept this idea. For him every thought emanating from his brain already presented some kind of truth, no matter how strange it might seem to others.

In 1908, the French physicist Duhem commented on the Galileo trial that scientific logic had actually been on the side of the Inquisition. He went as far as stating that, even if Copernicus’ hypothesis was able to explain all known phenomena, it could still only be concluded that it might be true, but not that it must necessarily be true. To do this, one would have to prove that there is no other system explaining the same phenomena as well or even better. The theory of relativity is one example.

The legend that the Inquisition had broken an old man, who at the end of his life would whisper on his death bed the words: “And yet it moves”, is indeed a legend and nothing else. There is no truth in it.

In 1976, the Austrian philosopher, Paul Feyerabend, who invented what is known as philosophical relativism, wrote that Galileo had not been a victim of medieval obscurantism. He pointed out that at that time the church had been much closer to reason than Galileo himself by taking into account the social and ethical consequences of the quarrel. Feyerabend felt that the final verdict was rational and just. The physicist Carl Friedrich von Weizsaecker went even further. His conclusion was that Galileo was travelling along the path that would lead directly to the atomic bomb, since Galileo was trying to advocate a science without any curbs at all. The Inquisition did not want to let that pass.

In the end, the Vatican conceded that there had been “mistakes” on their part at that time, that both the Pope and some members of the Curia did not support the outcome of the proceedings. Suddenly, for the church, Galileo became a man of high esteem with an exploratory urge, who regarded nature as the book of God. But it is still hard to counteract the myth surrounding this whole affair, which has accumulated over the centuries. It did not even help in 1741 when Pope Benedict XIV allowed the publication of Galileo’s complete works without any restriction.

The church’s opinion of the great man today is somewhat subdued. Cardinal Brandmueller, long time president of the Papal Committee for the Historical Sciences, still takes Galileo to be a vain self-obsessed scholar, who sometimes overdrew his account and certainly showed no scruples in the treatment of his colleagues and competitors. Brandmueller says the verdict had been well founded, firstly, because Galileo had obtained permission to publish the Dialogo by fraud, and secondly—once again—because he had been unwilling to present his theory as a mere hypothesis and not as the exact description of reality. He believes that the real scientific importance of Galileo lies in his last major work, laid down in the Discorsi, which he started to work on while in the "dungeon" in the palace of the Divine Office, while taking full advantage of the best cuisine in Rome.

As early as 1597, Galileo had confided in writing to a close colleague that he believed that Copernicus was right about Ptolemy's model. He also wrote about his views to Kepler in the same year, but insisted that he did not want to make a fool of himself in public and thus would not publicly support Copernicus' theory. His letter was a response to Kepler, who had sent him a draft of his Mysterium Comographicum.

He had obtained the book and letter through the good services of Kepler's friend Paul Hamberger. Since Hamberger had to leave Padua the same day, and Hamberger was due to carry Galileo's reply back to Graz as soon as possible, Galileo's reply on August 4th 1597 was just a courteous acknowledgment of Kepler's status of eminence as a renowned scientist, and did not delve into the subject matter of the book itself. He had only had time to read the preface. However, besides confessing to be a follower of Copernicus, he also explained why he had refrained from making any public pronouncements as such. He cited the fate of Copernicus. Although Copernicus had by no means been the target of persecution by the authorities, he had certainly been the subject of ridicule and derision. Galileo expressed the opinion that, as long as there was only a minority of scholars to follow the Copernican world model and a significant number of others still holding on to the old one, he was not prepared to go public himself.

One month later, in a letter addressed to Michael Maestlin, Kepler's friend and mentor, Kepler reported that the mathematician Galileo had been an adherer to the heliocentric system for many years and had requested two more copies of his work. He added a comment indicating that the book had resulted from the combined work of himself and Maestlin.

Kepler then forwarded the requested additional copies to Galileo on October 13th 1597, accompanied by a letter. The largest part of the letter was taken up by Kepler encouraging Galileo to come out into the open and boldly defend the Copernican model in public, while supposing that the prevalence of ignorance in Italy would be comparable to that in Germany. He did not go along with the idea of letting the unenlightened masses prevail over the truth, and refused to resort to ruses to convince other scientists. At the end of the letter he asked Galileo to carry out some observations of the pole star and Ursa Major in March 1598, since Kepler himself did not possess astronomical instruments of sufficient resolution.

Galileo addressed Kepler again in April 1611, using the good services of a certain Asdale in Prague to ask Kepler's opinion about the "Dianoia astronomica" of Francesco Sizzi, an Italian astronomer who had been the first to observe the evolution of sunspots. In this treatise, Sizzi refuted the existence of the Jovian satellites discovered by Galileo, who had published the results of his observations in "Sidereus Nuncius", on astrological grounds. The subtitle of the "Dianoia" continued as

"... where the rumor in the Sidereus Nuncius is proved to be unfounded". Kepler compared Sizzi to Martin Horky, another sceptic of Galileo's discoveries. His judgement of Sizzi was quite straightforward: he compared him to a blind man writing about the light of the Sun, but in the end indicated that he wanted to avoid harsh public criticism of the young author, and suggested he might befriend Sizzi in order to convince him.

All this led slowly but surely to Galileo's decision to opt once and for all for the Copernican world model. But in spite of his enlightened way of thinking, he continued to ignore facts that did not fit his own views, such as Kepler's discovery that the planets moved around on elliptical rather than circular orbits. Circular motion fitted a more traditional attitude towards nature.

Galileo admitted that there were phenomena in nature that might not be explainable, since human senses were limited. For example, he himself had no explanation for the substance of comets. This was the case in particular for the comet's tail. Sarsi declared that Kepler had refuted Galileo's claim that it was an optical illusion, which Galileo denied.

Thereafter, there was a long period of silence in the communication between Kepler and Galileo. Kepler's last letter had to wait another twelve years before Galileo would manifest himself anew, albeit without ever again referring to the Mysterium Cosmographicum.

Tycho Brahe

One other person to whom Kepler sent his Mysterium was the famous Tycho Brahe, then living and working in Prague as Imperial Mathematician of the Emperor Rudolph II. Brahe was fifty years old at the time.

Tycho Brahe proposed a world view that was a compromise between the heliocentric and the geocentric models (Fig. 5.3). During his lifetime, he was considered to be the "Prince of Mathematicians"—the greatest of all time! However, his model was rejected by Kepler—among others—on the grounds that it would destroy the integrity of the celestial spheres. It did not play a great role in science but rather served ideological purposes. After Newton, it disappeared completely from astronomy.

Brahe was born in Knudstrup in Denmark on December 24th 1546 (Julian calendar). Already early on in life he became fascinated by astronomy. On November 11th 1572, he detected a new star in the constellation of Cassiopeia, which made him famous. King Frederic II of Denmark left him the island Hven in the Oresund. There were 40 leaseholders on it, whose rents served as income to Brahe. It was here in 1576 that he built his famous observatory "Uraniborg", which became the centre of his astronomical research. Brahe carried out his observations for more than twenty years and collected much invaluable data. He also constructed a second observatory in Stjerneborg.

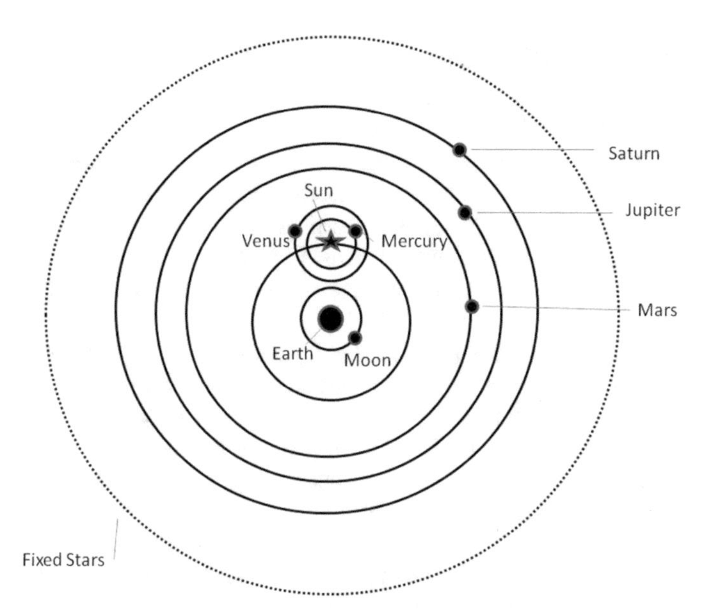

Fig. 5.3 Tycho Brahe's world model

However, he fell into disgrace and had to give up Uraniborg because of his ill-treatment of the leaseholders. He subsequently came under the protection of Count Rantzau in Wandsbeck, where he published his "Mechanics of the New Astronomy" in 1598. During that time Brahe had his first contact with Johannes Kepler.

In 1599, Tycho Brahe was appointed Imperial Mathematician by the Emperor Rudolf II, who provided him with Benatek Castle, southeast of Prague, as a place of work. His salary amounted to 3000 Guilders a year, an extraordinary sum in those days, to allow him to modify Benatek along the lines of Uraniborg.

Brahe went his own way with regard to Copernicus' ideas. He accepted the Copernican world model only partially, keeping the central position of the Earth just like Ptolemy, with the Sun and most of the planets going round it, but at the same time having the two innermost planets going round the Sun. This was therefore a kind of compromise between Ptolemy and Copernicus.

Concerning Kepler's approach to constructing his world model by geometrical means, Brahe wrote to Maestlin about it. He doubted that anyone would ever be able to build in an a priori manner upon the proportions of regular bodies something that would correspond to results obtained by pure observation. In this way, Brahe had put his finger on the problem with the Mysterium, and it became clear that any further research in this field would from then on be superfluous.

Bad Omens and the Counter-Reformation in Graz

Kepler married Barbara Mueller on April 27th 1597. After his marriage, his annual salary rose from 150 to 200 Guilders. Although they had two children, a boy, Heinrich, and a daughter, Susanne, both of them died a few weeks after birth (his wife had had a daughter from her first husband). The marriage was not really a happy one, since his wife was unable to follow Kepler's ideas. Kepler traced her melancholy state of mind back to an unfortunate configuration of Jupiter and Venus at the date of her birth.

Kepler was a superstitious person by today's standards. Because of his bodily weakness and multiple ailments, he was constantly looking for mysterious signs that would announce a new illness. For example, he thought he had discovered blood-colored cross-shaped marks on his left foot, which he compared to similar ones reported from Hungary, where they had been seen on human bodies, walls, and houses and been interpreted as omens for the advancing pestilence.

Meanwhile, in 1598, the Counter-Reformation gained in strength. In autumn 1598, the young Archduke Ferdinand, who had been educated by the Jesuits, ordered that all Protestant teachers of the monastery school were to leave the country under the threat of death, with the one exception of Johannes Kepler, who the authorities hoped would convert to Catholicism. His salary was even maintained but he was not allowed to continue teaching. At this stage Kepler remembered an invitation from Brahe in Wandsbeck to work with him in Benatek Castle, and he accepted.

However, the whole affair was not as straightforward as one might imagine. Just like today when we apply for a job, Kepler first had to go to Benatek and pass a test. Brahe assigned him the task of evaluating his Mars data. In addition, Kepler was not the only one in Brahe's entourage to be engaged in this endeavour, for there was competition from Brahe's assistant, Longomontanus.

From the very beginning, the relationship between Kepler and Brahe got off to a bad start. To begin with, Brahe liked to keep his Mars data close to his chest and gave Kepler only limited access. Then, in public, Brahe made no bones about the fact that he regarded Kepler as much inferior to himself. For instance, he placed his new assistant at the far end of the table whenever he took meals in the company of invited guests. In the week before Easter in 1600, it came to the crunch. Kepler exploded publicly about his master and his situation. Kepler left Benatek for Prague in a rage, staying there with his friend Baron Hoffmann. After calming down and reflecting on his material situation, Kepler offered a written apology to Brahe and subsequently returned to Benatek, after the latter had accepted it. From then on, Kepler's access to the Mars data was facilitated. Nevertheless, the relationship between the two astronomers remained one of mutual suspicion.

Kepler's resources had almost dwindled to nothing after about six months, with his wife and family still living in Graz, so he began to negotiate his future employment with the Emperor Rudolf II, once Brahe had confirmed that he still needed Kepler. At stake was a fixed salary for the next two years and travel expenses for himself and his family. While this was still going on, Brahe arranged for Kepler to join his relative Friedrich Rosenkrantz on a trip to Graz.

Kepler did not arrive too late. The archduke issued a new directive on July 17th according to which all inhabitants of Graz had to renounce the new faith and publicly convert to the faith of the ruler. Anyone refusing this act had to leave the country immediately. There was no exception for Kepler this time. He even had to pay a fine of 10 Thaler as a penalty for having had his first daughter Susanna baptised as a Protestant. The only concessions to him were a dismissal wage for half a year and the issue of a recommendation, which was full of praise about his engagement as a professor at his school.

Kepler communicated this development to Brahe, who responded by confirming that the emperor had accepted the conditions for Kepler's employment. He also informed him that Longomontanus had left him and relocated to Denmark. So, on September 30th, Kepler left Graz for good, with his wife and stepdaughter accompanied by two carriage-loads of household goods, and headed in the direction of Prague.

Chapter 6
Prague

1600–1612

Mars

One of the most important tasks assigned to Kepler by Tycho Brahe was, as already mentioned, the evaluation of his Mars data. This data concerned the positions and trajectories of that planet, not so much the characteristics of its surface or the composition of its atmosphere, which the instruments of that time would not have been able to measure. But it seems that, very early on in his observations of the celestial bodies in our Solar System, man developed a preference for the Earth's neighbour, Mars. This continues to this day. And there are now sound reasons for it.

Mars is the only planet we know where the conditions are such—albeit with the necessary support technologies—that humans could land on it and remain there for at least some limited time. There was even speculation that the planet was inhabited by Martians exhibiting similar characteristics to humans. This belief was sparked by observations made by the Italian astronomer Giovanni Schiaparelli in 1877, in which he claimed to identify canals on the surface of Mars.

Today we know the following facts about this planet:

Mars belongs to what are known as the Earth-like planets, i.e., planets not far from their star with a small diameter and a small mass, but high density, possessing a tenuous atmosphere consisting mainly of nitrogen, oxygen, carbon dioxide, and some noble gases. Their internal composition is basically stone and metals, and they may or may not have one or two satellites.

Mars is the second smallest planet in our Solar System. At the same time, it is the one that most closely resembles the Earth, although its mass is only about a tenth that of the Earth. The gravitational acceleration is 3.69 ms^{-2}. When compared with the Earth, its axis of rotation has a similar inclination to the plane of its orbit, and so there are also seasons on Mars. Its highest temperature is a comfortable 27 °C, though the mean temperature only −55 °C. Ninety-six percent of the atmosphere is carbon dioxide with small traces of nitrogen, argon, and oxygen. Probes have found embedded water ice at the polar caps.

© Springer Nature Switzerland AG 2020

W. Osterhage, *Johannes Kepler*, Springer Biographies,
https://doi.org/10.1007/978-3-030-46858-3_6

Its surface displays ditch-like structures and volcanoes, river valleys, indicating past water floods, and desert-like areas. Mars has two small satellites: Phobos with a diameter of 27 km and Deimos with a diameter of only 18 km, both being irregular pieces of rock.

Mars has been the target for a vast number of earthly visitors in the form of space probes, robots, and automated vehicles driving around on its surface. The first two vehicles to reach the surface of Mars were the landers of the Soviet probes Mars 2 and 3. Both were destroyed during or shortly after touchdown. These attempts were followed by the successful Viking Missions of NASA and more than a dozen others involving half a dozen countries, culminating in the USA's Mars Science Laboratory, still in operation. One of their recent objectives is to find a suitable landing site for human visitors in the not so distant future.

Prague

Prague is situated in the north-west of Bohemia on the Vltava River (Fig. 6.1). At the times we are considering here, it was the capital of Bohemia and home to the Holy Roman Emperors, notably Rudolph II. Rudolph II had philanthropic leanings and invited an illustrious assembly of people to Prague Castle, where he lived. Some of these were famous, others not so: scientists, musicians, artists, and magicians of all sorts. Among the most famous were Tycho Brahe and later, through the association with Brahe, Johannes Kepler.

Tycho Brahe engaged Kepler permanently and paid him from his own salary. The first task he gave his assistant was to draft a pamphlet against Brahe's deceased predecessor, the mathematician Ursus. This Ursus had tried to sell Brahe's world model as his own. Unfortunately, Kepler had dedicated his Mysterium Cosmographicum

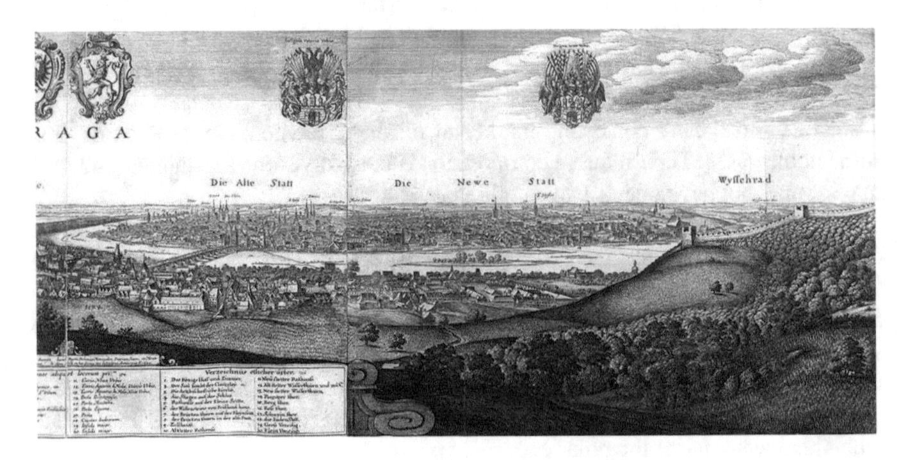

Fig. 6.1 Prague in 1650, by Merian

to Ursus. So this was a double revenge by Brahe. The relationship between Brahe and Kepler turned sour again, leading to Kepler's leaving the vicinity of Prague after various quarrels with his master. He had to return to Graz to solve a number of financial problems which had arisen after the death of his father-in-law. In August 1601, Kepler returned to Prague to continue working with Brahe for another two months. The two astronomers made observations from the balconies of Queen Anna's summer residence, Belvedere.

Shortly afterwards, Tycho Brahe fell ill with a bladder ailment and died on October 24th 1601. Johannes Kepler was at his death bed in Prague. Brahe was buried in the Utraquistical Tyn Church, and for that occasion Kepler composed a long eulogy of 184 verses praising his master and Emperor Rudolph (Fig. 6.2). But now at last Johannes Kepler had unrestricted access to Brahe's data, because only two days after Brahe's death the emperor had engaged him as the official successor of the Danish astronomer, albeit for an annual salary amounting to 500 Guilders instead of the 3000 Brahe had been paid. And even these 500 Guilders were disbursed only sporadically, partially, and in the end not at all. Among his duties as a court mathematician, Kepler was also asked to produce horoscope s, whenever such requests arose. At the same time, this activity presented an additional source of income for him.

Kepler's main inheritance from Brahe were twenty-four folios of raw data from Brahe's observations. They would serve him to calculate exact sky maps for all known planets. The final result would be the famous Tabulae Rudolphinae. Brahe himself made his inheritors pledge to edit and finalize them. It would take nearly twenty years to fulfil this obligation, and by then, the Emperor Rudolph was long dead. During his lifetime, Brahe had not managed to solve the riddle of the planetary

Fig. 6.2 Emperor Rudolph II, painted by Joseph Heintz the Elder in 1594

orbits. It did not help Brahe to possess the most refined instruments and a staff of qualified assistants, when he had no notion of the master plan behind the structure of the Universe he spent all his time observing.

But before Kepler could finalize his own theories, other matters required his attention. To prove his aptitudes as a mathematician, he produced an update of the works of Erasmus Vitellio, which went as far back as 1270. Vitellio, born in Silesia, had compiled a work about solar and lunar eclipses and at the same time about the diffraction of light, especially in the atmosphere. In 1603, Kepler published a volume called "Ad Vitellionem Paralipomena, quibus Astronomiae Pars Optica traditur" (Addendum to Vitellio and Optical Part of Astronomy). In the New Year of 1604, he handed his manuscript over to the Emperor. Now he could turn back to his main interest: the astronomical master plan. This would take another four years—until 1609. The title of this most important publication reads:

"Astronomia Nova αιτιολογετοσ seu physica coelestis, tradita commentaries de motibus stellae Martis ex observationibus G. V. Tychonis Brahe" (New Astronomy, Based upon Causes, or Celestial Physics, Treated by Means of Commentaries on the Motions of the Star Mars, from the Observations of Tycho Brahe).

The title indicates two important features:

1. The work does not limit itself to just describing observations and structures, but aims to explain causes by resorting to something new: celestial physics, as yet unheard of in such a context.
2. It is based solely on the Mars data provided by Brahe and evaluated by Kepler over a period of ten years.

What does "Celestial Physics" refer to and why did Kepler introduce it in the first place? We shall come to some answers later in this chapter. However, the need arose when Kepler assumed for the first time in astronomy that the paths of the planets were the result of their motions in free space rather than being due to their attachment to rotating spheres. One consequence was that he had to introduce a new mathematics as well, and of course this differed from the mathematics that would be used to describe uniformly moving ether spheres. Indeed, it had to take into account natural forces causing the irregular and free motion of the planets. As a result, this new astronomy based on mathematical hypotheses differed considerably from Aristotelian physical astronomy.

Force

Prosaically, we represent a force by an arrow and denote it by the capital letter "F" for "force". But before arriving there, we have to take a few steps backwards, because the notion of force is associated with something mystical. To quote H. Bohn's "Leitfaden der Physik" of 1915 [17] (translated by the author):

> In the year 1666, Newton had the idea that the central force causing the Moon to circle the Earth would be gravity [...]. The forces act on one another in inverse proportion to the

> squares of their distances [...]. This law has been proven with respect to the motion of the Earth and the other planets around the Sun; it is valid for minor motions on the Earth as well as in space as a whole.

The mystical aspect is contained in the word "force" itself. Even though the above statement sounds rational, it is shrouded in terms which were coined in non-scientific epochs. Regarding its original meaning, a "force" had always been something mysterious, quite often something threatening from afar, emanating from powers that might or might not be appeased. Natural science later took over this term from its irrational origins, using it to formalize the observed conditions in the world and make it into one of its foundations. That may also be one reason for the complexity of the resulting systems of equations. Later, Einstein was one of many who tried to get rid of "force" once and for all by replacing it with geometrical structures. But since our reasoning has got used to the good old mysticism, his approach actually caused even more headaches for most of us.

Expressing the definition in the old physics book in simpler terms:

The environment in which a force acts is called a field of force. And the first classical field of force to be described mathematically was the field in which the gravitational force acts. As mentioned above, gravitation.

Further down in the same textbook, another force is introduced:

> In the years 1770–1780 Coulomb found by very careful measurements that the strength of the force by which two magnetic poles would act upon one another depends on the inverse square of the distance between those two poles.

This sounds very similar to Newton's observations of gravity. Such formal analogies not only possess some anecdotal charm, but have led people again and again to look for some common substance behind the formal similarities. Later in this book (Chap. 10), we will discuss such unification theories and also other forces dominating our Universe.

We have already mentioned that forces create effects. The long range effect of forces is of particular interest. In the case of gravitation, masses of compact bodies located at a certain distance from one another interact across this distance. In this case the effect is always an attractive one, whereas in electromagnetism it could also be repulsive.

Gravitation

For a long time, scientists adhered to the theory of the long range effect of gravitation: a body is induced to move or induces another body to move by the action at a distance caused by gravitation. The reaction of the partner in this game seems to occur without any delay in time. However, this leads to theoretical problems. If this instantaneous action is a fact, and observation seems to support it, this would mean that the reaction of a test body must happen simultaneously with any action or motion of the body in whose gravitational field it is traveling—although it may be at some considerable distance from it. Thus the transfer of the effect of a gravitational force would have to occur at infinite speed. This is in glaring contradiction with the findings of the special theory of relativity. The latter says, among other things, that there is a maximum speed in nature, which is constant in all directions of motion: the speed of light. And it possesses a finite value.

If we assume that the speed of transfer of the effects of gravitation is equally finite—at most the speed of light—we stumble upon another dilemma: during the energy transfer from body A to body B, the energy would cease to exist until it arrived at the target body. This contradicts the law of energy conservation.

To solve these problems, the assumption of the long range effect has been replaced by the theory of the close range effect. Another observation about fields of force is that a field shows an effect only when particles or test bodies travel through them. Only then does the nature of the field become apparent and only then can its forces be measured. Thus the first body is surrounded by a field of force. The test body causes a disturbance within the force field in question whenever it enters it. The whole event takes place in the space surrounding the first body, in such a way that space itself becomes a medium endowed with physical properties. Thus force effects are conveyed by space. Long range effects have been replaced by close range effects between the test body and the field surrounding the original body.

In this way, space acquired a different quality. Although it was originally only of use as a basis for geometric reference, it could now actually take part in physical events. This would take on a much greater significance than in classical physics when the general theory of relativity came along.

Astronomia Nova

The work is organised into five parts with altogether seventy chapters. It commences with a dedication to the Emperor Rudoph and a general introduction. However, Kepler was forced to accept an additional preamble by Frans Tengnagel, a son-in-law of Brahe who administered Brahe's heritage, including the data for the Tabulae Rudolphinae, which Kepler refused to hand over to him. As a compromise, Tengnagel was allowed to compose this additional introduction, which was, however, completely irrelevant to the body of the work itself. Tengnagel later became a corresponding member of the Accademia dei Lincei, founded by Federico Cesi, of which Galileo was also a member.

The five parts of the Astronomia covered the subject roughly as follows:

Part I: Comparison of existing planetary theories:
What would be the effect on the Mars trajectory if the Sun were introduced as the centre of the world rather than the Earth?

Part II: Discrepancies between the motion of the planet Mars and the old hypotheses:
Critical investigations of the observations by Tycho Brahe regarding the Mars orbit. Deduction of a new hypothesis.

Part III: Further discrepancies with regard to the motions of the Sun and the Earth: Investigation of the relative motions of the Sun and the Earth as the key to a new astronomy. Calculation of the distance between the Sun and the Earth. Proof that the speed of Mars is inversely proportional to its distance from the Sun. Physical investigations about the force exerted by the Sun on the planet and the planet's own contribution, together with the associated equations.

Part IV: Discovery of the elliptical orbit s of Mars and the other planets:
Conclusion that the Mars orbit is not circular. Calculations of various distances between Mars and the Sun. Kepler's first law: the orbit of Mars (or any other planet) is elliptical with the Sun at one focus.

Part V: Details of the Mars orbit and criticism of Ptolemy:
Inclination and parallax of the Mars orbit.
Most importantly Johannes Kepler rejected all three competing world systems of the day: Ptolemy's, Copernicus', and Brahe's. Kepler claimed that the three models were in fact indistinguishable from one another, since all three could be tailored in such as way as to fit the corresponding calculated results with observable data. Mathematically and kinematically, all three competing world systems were compatible. However, this was true only for short-term observations. In comparison to historical data, all three showed deviations.

The salient details of his work can be summarized in the following way. Copernicus had postulated that the centre of the Universe would be a point near the Sun. This was rejected by Kepler, who placed the Sun itself at this point. At the same time, the Sun became the mover of the planets. This is where he approached the question of gravitation for the first time. He also determined that the planets were not moving at uniform speeds, but rather at speeds that varied according to their distance from the Sun (Kepler's second law). This led him finally to the already mentioned conclusion that the Sun must be the cause for the planetary motions. In his terms, the Sun must emit some agent, similar to light, which would guide the planet, whereas the planet itself would have to exert some additional force itself to keep it from drifting into outer space. Kepler compared this attractive force to magnetism. He speculated that it was proportional to the size of the heavenly body in question, but that it had only a limited range.

There were two further reasons why, at the outset, out of the two models available in around 1600, Kepler preferred the Copernican model over the geocentric Ptolemaic picture. One reason was that he could systematize all planetary motions in one closed system without having to think about the components of the apparent motion of each individual planet due solely to the motion of the Earth. And the second was that he would be in a position to determine the absolute distances of all planets from the Sun empirically. These considerations provided plausible arguments but could not be proven, even when empirical data were consulted.

But if the heliocentric model did describe the way things really were, it had to be derivable from real causes—causes which had ontologically to be of a higher order with respect to divine creation. They had to be the blueprint for creation itself. This was the key Kepler was looking for. And it had to be found in geometry, since he assumed that geometry would be the building block on which God had created the Universe. Kepler's firm belief was that God never did anything haphazardly. In his Mysterium Cosmographicum, Kepler had already noted that his relentless search for the real causes involved the number, size, and speed of the planetary spheres.

In the end, Kepler took the improved observational data from Tycho Brahe to further confirm his Mysterium Cosmographicum. This data led him to the phenomenon of orbital periods increasing in proportion to the radius of the corresponding orbit. His conclusion was that the cause for this motion was to be found in the central body—the Sun. The effect of this cause, however, decreased with the distance. Why this should be so remained a mystery to him, only to be solved later by Isaac Newton.

Another consequence of his reasoning was to abandon the concept of ethereal spheres altogether. This was a decisive step away from Aristotelian celestial physics. Kepler was seeking an astronomy quite different from that of the ethereal spheres, because it would relate all motion to a single force. Initially, he arrived at this conclusion solely by deductive reasoning. However, instead of stumbling upon gravitation, he found the cause in a kind of cosmic magnetism, as proposed by William Gilbert in 1600.

Concerning the elliptical paths of the planets, Kepler initially tried to construct a model that was once again based on epicycles, which had been the basis for Ptolemy's pre-Copernican geocentric model, but he could not find a physical explanation for it and thus discarded this approach. Instead, he sought another explanation, assuming some sort of spiritual disposition of the planets which he tried to combine with Gilbert's theory of cosmic magnetism. Sticking to this kind of reasoning, he missed the possibility of considering mutual attraction between two celestial bodies based on the existence of a single force—gravitation.

As indicated by his title, Kepler's findings were originally related to the trajectory of Mars, but he extended his conclusions to the other planets as well. However, he was not satisfied with his earlier geometrical model presented in the Mysterium Cosmographicum, even though it seemed to have been confirmed by Brahe's data base. His polyhedrons could only approximately determine the relationships between the planets. The cosmic order and its harmony had to be found elsewhere.

For Kepler, the key was the harmonic ratio of the extreme orbital speeds of two neighboring planets. This even holds true for planets found after Kepler, although there is still no plausible explanation in modern science. By taking this approach, Kepler went back to the Platonic idea of a grandiose overall view of the cosmos, comprising music, astrology, the human soul, human society, and the paths of the planets. He endeavored to create a numerical astronomy by harmonizing physics with other disciplines. His search for the underlying causes explaining how his numbers came about was not an end in itself. In this he differed for example from Galileo. Kepler wanted to discover the plan of creation itself as a means to find God and thereby praise the Lord [18].

In any case, from a scientific point of view, Johannes Kepler was a true pioneer of astrophysics, since he—in contrast to other astronomers before him, for whom the motion of celestial bodies was a mere kinematical problem—had discovered the causal role of the Sun in these motions.

New Discoveries

It was in the same year 1610 that Kepler got into contact again with Galileo, when he learned that the Italian had discovered the Jovian satellites. This discovery worried him to the extreme, since it had the potential to destroy his just completed work on world harmony. Through the good services of the Tuscan legate in Prague, he obtained a copy of the Siderius Nuntius, in which Galileo had published his findings, together with Galileo's verbal request for Kepler's comments on it. Kepler replied with a paper titled "Dissertatio cum nuncio sidereo" (dissertation on the Siderius Nuntius). In his reply, Kepler took Galileo's statements at face value, without having been able to check the scientific facts. Galileo appreciated Kepler's paper as an important piece of support in his quarrels with his colleagues and adversaries at that time.

Kepler's support for Galileo was not completely without criticism, and this led him to request the names of witnesses who might have seen the celestial phenomena. This was four months after the dissertation, which Galileo had not even acknowledged by then. In his reply, Galileo cited the Archduke of Tuscany as witness, rather than a scientific peer. He also declined to lend his telescope to Kepler and said he was not in a position to produce a duplicate. This episode concluded the sparse direct communication between the two renowned astronomers. However, Galileo continued indirectly, via the Tuscan legate, to bombard Kepler with a series of anagrams, leaving Kepler the task of decoding their meaning, about new discoveries around Saturn. Galileo did not identify Saturn's rings as such, but interpreted what he saw as satellites residing on opposite sides of the planet.

In the end Kepler borrowed a telescope from the royal household and confirmed by the existence of the Jovian satellites his own observations. The resulting publication was called "Narration de Jovis satellitibus" (statement about the Jovian satellites). It was printed in Florence, from where it reached Galileo and helped to improve the latter's credibility.

Telescopes

The instrument that was decisive in confirming or disproving any astronomical theory was the telescope. It was probably invented by the German spectacle-maker Hans Lippershey (1570–1619), although even he may not have been the first. He looked through a tube "to see things far away as if they were nearby". In fact he did not obtain the patent for his telescope, which he applied for in 1608, because another member of his trade had already filed for one for a similar invention at around the same time.

Lippershey was born in Germany near the German-Dutch border, but lived most of his life until his death in the Netherlands and became a Dutch citizen. There exist several different stories about how he came to the idea of building a telescope, also referred to as the "Dutch perspective glass". According to one report, he just simply copied one that had been invented earlier by someone else. In any case, his telescope used either two convex lenses or one convex objective and a concave eyepiece. The latter version produced an upright image for the viewer and its magnification was 3X.

Soon after Lippershey's patent application, the design of his instrument was disseminated all over Europe, and this brought others to improve it. One of them was the Englishman Thomas Harriot (1560–1621). This astronomer was the first person to look at the Moon through a telescope and to draw a picture of the resulting slightly enlarged heavenly body—even before Galileo did so. He also detected sunspot s one year later, in 1610. Just as the invention of movable letters in printing came at just the right time to further the dissemination of Luther's reformation, so the invention of the telescope came at just the right time to substantiate the Copernican world model.

In the end, it was Galileo who first perfected the telescope. He constructed it from lenses he could buy on the market. Initially, its magnification was 4X, but later this rose to 8X and even 33X. He demonstrated its use to the government of Venice on the 25th of August 1609. The members of the government were deeply impressed. Galileo's prime use of the instrument was the observation of heavenly bodies. But he had previously written to Leonardo Donato, the Doge of Venice, to extol the virtues of his invention in quite another context, calculated to arouse the interest of a man of power who generally had other designs than astronomy. On the 24th of August, he had explained that it could be useful to any business on land or water, such as spotting enemy ships much earlier, when they were approaching from afar—about two hours earlier than without a telescope. This would enable someone to judge the size of the enemy fleet and its capabilities and take early action to prepare for its defeat. On land, his instrument could be used to spy on enemy fortifications and movements.

The device had a collecting lens as objective and a diffuser lens of short focal length as eyepiece. The focal points of the objective and eyepiece coincided on the observer's side. It had only a small visual field, but represented objects in an upright position and not reversed. Today, this configuration is still used in opera glasses. Since the ocular has a negative focal length, it has to be positioned within the focal length of the objective. No real intermediate image is produced. The advantages of the Galilean telescope are to be found in its short length and its upright image. Its disadvantages are the narrow visual field and—compared with Kepler's telescope—the difficulty in locating observed objects precisely, since it was not possible to use cross hairs due to the absence of any intermediate real image.

Johannes Kepler presented the design of his telescope in his publication "Dioptice" in 1611 (Fig. 6.2). His instrument consisted of an objective and an ocular, both convex. As a result he initially got an inverted image of an observed object. However, it could be made upright again by employing a third convex lens (Fig. 6.3).

Concerning the history of the telescope, however, Paolo Galluzzi told another version of the tale in "The Lynx and the Telescope" [19]. According to him, Federico Cesi, the founder of the Accademia dei Lincei, decided in 1612 to reconstruct the history of the invention of the telecope on the basis of testimonies from members of the Accademia. He came to the conclusion that the theoretical framework for the

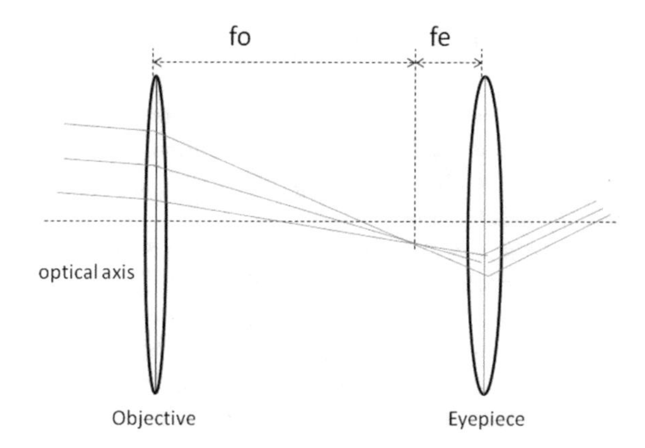

Fig. 6.3 Optical paths in Kepler's telescope

instrument had been developed by his co-founder Giovan Battista Della Porta as early as 1609, whereas Galileo worked in parallel to refine it. This version had been endorsed by Kepler himself in his correspondence on the Siderius Nuntius, giving Galileo full credit as the main inventor. Kepler dismissed Lippershey's achievement as a matter of pure luck, with no scientific foundation. But still, Galileo remained unhappy about the assertion that Della Porta should have developed the optical theory before him, since he claimed that part of the story for himself as well.

Cosmic Harmony and the Mind

The fascination with spheres and their perfection continues well into our time. The German mathematician Guenter Roeschert was inspired by the concept of the internal and external topology of the sphere [20], just like Kepler. In a thought experiment he places a sphere of infinite size, representing the cosmos, into a four-dimensional manifold. He then distinguishes three infinities describing this entity: the external infinity outside the sphere, the internal infinity inside the sphere and the infinity of the surface of the sphere. A surveyor or observer may place himself at the centre of the sphere, although there is no mathematical necessity for this.

Without going into the details, Roeschert uses this abstract construct for the world to obtain, what he calls a vision transcending "the abyss of the world"—referring to what we have called a perceived chaos at the beginning of this book. His conclusion is that the ordinary mind transforms such an inconspicuous object as a sphere into a perfect description of reality.

Endgame in Prague

Bad omens once again signaled the beginning of the end of Johannes Kepler's engagement in Prague. It was the year 1611. First, his son Friedrich fell ill with smallpox and died. Then his wife Barbara also died, having contracted the same disease from the boy shortly afterwards.

The Emperor Rudoph's position was under ever greater threat from the designs of his brother Matthias. Rudolph abdicated on May 23rd and Matthias became Emperor on June 13th. Owing to the political turmoil, Kepler was unable to continue his scientific work. However, the abdicated Rudoph convinced the astronomer to stay with him until his death on January 20th 1612. Kepler was then free to move. He entrusted his two surviving children to a widow acquaintance in Kunstadt in Moravia, and a year later to another family in Wels in Upper Austria. He arrived in Linz as a single man in May 1612.

Chapter 7
Linz

1612–1626

Science and Theology

The ambivalent relationship between science and theology was not a topic we may relegate solely to the onset of modern times when Johannes Kepler was struggling with it. It continued far beyond that era and well into the present. An example can be found in a letter written by the student of theology Ernst Eduard Kummer (1810–1893) to his mother. He wanted to quit theology for something different:

> Do not believe that I am surrounded by anxious doubts. No, it never stood more clearly before my soul that a person should do right under any circumstances, without looking for any reward at all. But I do not consider profane happiness to be the highest asset of mankind; I prefer rather the peace of mind which is derived from the conviction of having acted justly. As long as I have this conviction, I shall not be tempted by a base displeasure to despair of God and immortality, even if I should have realized by rational means that the spirit is immortal and that there is a God, who has called this spirit into being, not to destroy it, but to elevate it to its highest perfection, which will be its bliss. Now, with a clear conscience, I cannot continue to study theology, and so I have abandoned it and have chosen mathematics, because it is a science in which anyone who reaches deeply enough will not be misunderstood by others or taken to be godless or evil, but in which one can find a truth which has to and will be recognized by all. [21]

Kummer chose mathematics because he thought it to be ideologically neutral, with no quarrel about heresy between theologians. However, that was not the way Kepler tackled the problem in his time.

Truth

There has been abundant discussion about the law of excluded middle (tertium non datur), which suggests that a proposition may only be "true" or "false" and nothing in between. Mathematicians apply this concept for example to the conjecture by Christian Goldbach (1690–1764) that every number greater than 2 can be expressed

© Springer Nature Switzerland AG 2020

W. Osterhage, *Johannes Kepler*, Springer Biographies,
https://doi.org/10.1007/978-3-030-46858-3_7

as the sum of two prime numbers—neither proven nor disproven until today. One fervent follower of the law of excluded middle was David Hilbert (1862–1943), who stated in a lecture in Bologna in 1928:

> How would it be with regard to the truth of our knowledge in general and to the existence and the progress of science if there were not even any certain truths in mathematics? [...] In mathematics, there is no Ignorabimus. We can always answer meaningful questions, and thus it is confirmed, as Aristotle had perhaps already sensed, that our mind does not ferment some arcane arts, but is rather structured along completely arrangeable rules and thus is at the same time the guarantee of the absolute objectivity of its judgment.

The law of excluded middle found its way into Boolean logic as well, and plays an important role in computation. But long before it was formalized, it underpinned Kepler's reasoning and final objective. Indeed, in science, and in particular in astronomy, this assumption was for him the only one possible: there could be only one truth, whereas in theology he remained ambivalent about which of the three systems of thought—Catholicism, Lutheranism, or Calvinism—would turn out to be the true one, as we shall see later in this chapter.

Incidentally, the question of the law of excluded middle in mathematics and logic is by no means decided, as can be seen from Osterhage, "Mathematical Theory of Advanced Computing" [22].

Linz

Linz is the capital of Upper Austria, situated on the Danube just north of the foothills of the Alps (Fig. 7.1). Its foundation dates back to the Celts in about 400 BC. It was subsequently upgraded to a fortified castle by the Romans. In the early Middle Ages, Linz developed to become a proper town, and in the 13th century the seat of the governor. The Emperor Friedrich III took up residence there from 1458 to 1462, and so the town became the center of the Holy Roman Empire for a brief period. During the Reformation, Linz initially became Protestant, but from 1600 onwards, the Counter-Reformation gained ground. That was the time when Johannes Kepler was in residence there.

Linz became famous again in modern times, when Adolf Hitler went to school there. Hitler had great designs in mind after the "Anschluss", when he annexed Austria to Germany. Linz became an industrial center, focusing on armaments (the Hermann Goering plant—today Voest Alpine) and chemistry (Stickstoffwerke Ostmark—later Chemie Linz).

Linz was where Johannes Kepler stayed for the longest consecutive period of his life—fourteen years altogether. He gave lessons in mathematics and Latin at a small Protestant high school, founded around 1550, a position created for him by well-meaning patrons. They were practically the only ones who realized the importance of their protégé in this rather provincial town. The school was organized according

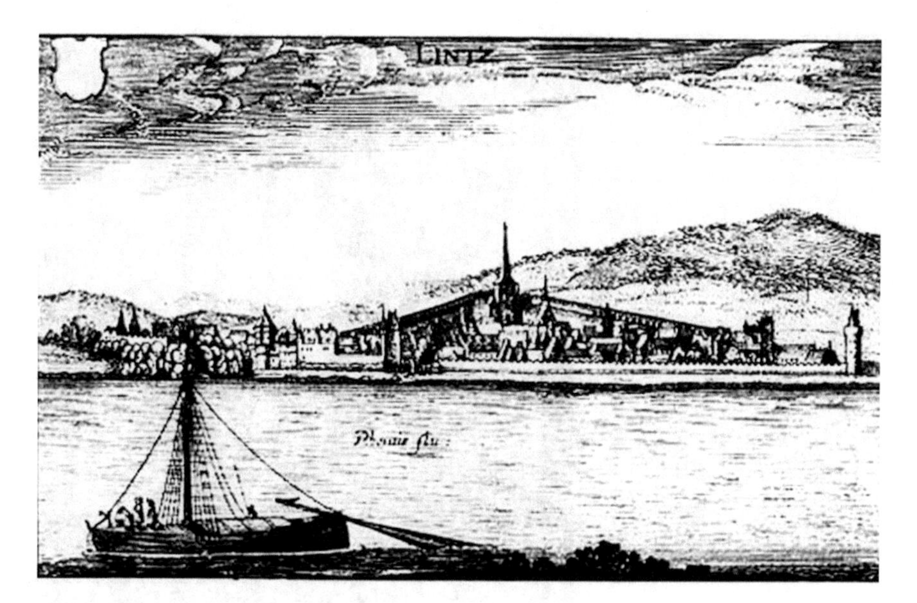

Fig. 7.1 Historical view of Linz by Merian

to the school regulations developed by the reformer Philipp Melanchthon in Saxony. The schoolmaster was instructed to adhere strictly to the Confessio Augustana, and to avoid any sects.

Bernegger

In July 1612, Kepler made the acquaintance of a new friend: Matthias Bernegger (Fig. 7.2), who was on his way to Strasbourg. Although they only met once, they continued their friendship by letter until shortly before Kepler's death. Bernegger himself was well acquainted with Kepler's achievements. He was ten years younger than the astronomer, born in Hallstatt in the Salzkammergut in 1582. He was a famous philologist and a professor in Strasbourg. In the 1630s, he visited and corresponded with Galileo, and organized the publication of the Italian's works from Volgare into Latin.

In 1613, Kepler married for the second time. His wife, Susanne Reuttinger, was a burgher's daughter from Eferding in Upper Austria. Together they had six children, of whom three died at a very early age.

Fig. 7.2 Matthias Bernegger

Point of Origin and Chronology

Many centuries after Kepler, in 1968, the German theologian Dietrich Wiederkehr wrote: “In Jesus Christ a fact was created by which everything past and everything later could be related to a central point.” [23]

It was to this central point of origin in our space-time coordinate system that Johannes Kepler next turned his attention. Wiederkehr identified this point in time with the crucifixion of Christ, whereas Kepler was looking for a point in time some thirty years earlier: the year of Christ’s birth. As his first task in Linz, Kepler had temporarily laid aside his astronomical interests and devoted himself to chronology.

In the year 525, the Monk Dionysius Exiguus presented his Christian Chronology in Rome. It was a kind of byproduct of his efforts to calculate the Easter cycle. In 325, the council of Nicene had decreed for theological reasons that Easter should always be on the first Sunday after the first full moon after the spring equinox. This rule comprises thus two astronomical reference points:

- the full moon and
- the spring equinox.

Furthermore, the seven day week is not an astronomical quantity. Since then a number of refined calculations have been developed. The last attempt came from Carl Friedrich Gauss with a correction algorithm in 1816.

Exiguus based his calculations on the so called Alexandrian Cycle, which combined the 19 year lunar cycle with the 28 year solar cycle. Until his time, chronology had been based on years after Diocletian's rule. Exiguus completed his Easter tables with the additional reference column: "anni ab icarnatione Domini" (years after the incarnation of the Lord), starting on the 1st of January 533, which corresponded to 248 after Diocletian. From Exiguus' tables, other people calculated the birth of Christ (1 AC, "0" was not known as a number then) and found it to correspond to the year 754 after the foundation of Rome. It took another five hundred years before Christian chronology gained ground in public and official documents, since princes continued to count the time in years after their accession to the throne.

Exiguus' chronology did not remain without critical reaction. In 1991, Heribert Illig, a German publicist and editor, published his theory that 297 years of written history did not really take place between 614 and 911. He called this period the "invented Middle Ages" and proposed to adjust existing chronology. Serious scholars later refuted his ideas.

Kepler, however, in his Komputistik, arrived at the conclusion that King Herod had already died four years before the turning point in history. This meant that the birth of Christ must have taken place in the year 5 before our common era. Kepler released his manuscript to be printed in 1613 through the good services of his new friend Bernegger in Strasbourg. The original title in Latin read:

"De vero anno natali Christi" (The True Year of the Birth of Christ).

In reality the exact year of Christ's birth is unknown. Even his contemporaries did not know it, and the narratives in the gospels are contradictory and unhistorical. Today there is a consensus that Christ was born between 7 and 4 BC. An additional counting problem arises because of the fact that the year 0 does not exist, since the number 0 was not in use in Europe until the end of the Middle Ages.

The Heretic

In Prague, Kepler was left in peace with regard to questions concerning his faith, but in Linz history repeated itself compared to his years in Graz. At the centre of a major dispute was a man who was both a compatriot from Wurttemberg and also a former scholarship holder of the duke, the clergyman Daniel Hitzler. Hitzler was a school superintendent. Otherwise, he was a person of little importance. The conflict—provoked by Hitzler—centered on what was known as the formula of concord.

To understand the conflict and its context, it is important to bear in mind that, at the time of Kepler's appearance, theology was all about disputes with the adversaries of other confessions. At the centre of the arguments was concern about salvation. In addition to theological contradictions, political circumstances dominated the agenda, since faith was bound to the confession of the respective ruler. This was the spiritual environment in which Kepler had to come to grips with his own convictions and the boundary conditions dominating his existence.

Let us step back from Linz to the very first roots of the conflict. Everything started with Kepler's baptism. When he was born, a Protestant baptism was not possible in Weil der Stadt, so it has to be assumed that he was baptized a Catholic. When in later years the Jesuits wanted him to convert to Catholicism he replied that he had actually never left the Catholic church in the first place, but—and this is the salient point—he had always been instructed by the true teachings of the church—referring, however, to the Confessio Augustana. So Kepler had been educated according to Lutheran teachings.

Very early in his youth, at the age of about twelve, he started to worry about theological problems. These were questions relating to the Last Supper and the person of Christ. Later, the doctrine of predestination and Luther's unfree will became of great interest to him.

First, serious disputes about the doctrines of the Last Supper had arisen in Adelberg between Kepler and the young preacher community there. The controversy centered on the confutation of Zwingli's interpretation. For Zwingli, the Last Supper as celebrated in the church was only a repast of remembrance with some symbolic power. For Kepler himself, the discussion led directly to the understanding of the two natures of Christ. In vain he tried to mediate between the Calvinist and Lutheran positions. Later, in Maulbronn, his doubts increased and he became increasingly stressed by the whole issue. He found it more and more difficult to put with with the condemnations of people belonging to one or the other confession.

While working in Graz as a professor of mathematics, Johannes Kepler wanted to return to Wurttemberg to serve as a minister, but nobody in Tubingen had any interest in calling him back as a theologian. His theological positions did not conform to the ideas of the Lutheran confession in Wurttemberg. With his Mysterium Cosmographicum, he turned away from his theological ambitions for good, and started to study the book of nature as the revelation of God's creativity. We know little about the details of Kepler's persuasions then, although there exists an exchange of letters between him and a friend, Colmann Zehetmair, in which Zehetmair referred to an essay by Kepler, dated to 1599, regarding the latter's conception of the Last Supper. In this essay, Kepler apparently doubted the presence of the substance of Christ's body, but accepted the dedication of the fruit and the merits of his death in the celebration. Kepler began to perceive the real presence as something absurd. This brought him somewhat closer to Calvinism. In his interpretation, Calvinists and Jesuits were much closer to one another than the two parties would admit, since both derived their interpretation from the Church Fathers and the scholastics.

In Prague, Kepler's theological considerations were still being turned over in his own mind, but did not lead to any external conflicts. His three children born in Prague were all baptized by ultraquist clerics, whereas the funeral address for his wife was given by Matthias Hoe von Honegg, an influential Lutheran and sharp critic of Calvinism. He had become a member of the directorate of the German Church in Prague and tried to persuade Kepler to give up his Calvinist leanings. But Kepler's main concerns at that time had to do more with questions about the spirits responsible for the motion of comets, which Kepler solved by deciding that the Sun was the central mover of the world. Another subject of discussion concerned the usefulness of a scientific astrology.

In an exchange of letters with Johannes Pistorius the Younger, a German medic and Catholic theologian who had converted from Lutheranism, Kepler was forced to align himself with Catholicism. But in a letter dated June 15th 1607, Kepler criticized the Catholic hierarchy as having deviated from the exhortations of Christ and the apostles in favor of the vanities and vacuity of the world. At the same time, he asked his correspondent to destroy this openly critical letter, and apparently, this did not happen. However, Pistorius replied that Kepler had no notion of theology at all and should return to his mathematics. Which Kepler did.

In 1610, he addressed a letter to Hafenreffer announcing that he was working on a "conceptum Germanicum", in which he would try to pave the way for peace between the confessions. While still in Prague, in 1609, Kepler had begun an exchange of letters with representatives of the Lutheran Church in Wurttemberg, especially with his old acquaintance Hafenreffer, and this for two reasons. First of all, Kepler sensed that his position in Prague was not secure, due to the political turmoil. So he inquired about the possibility of employment in his former homeland. But people there were less than keen to have him around again. And this is in connection with the second reason: Kepler's stance with regard to the meaning of the Last Supper and his interpretation of Christology. On both accounts, Kepler saw himself in no position to sign the formula of concord, or only with the proviso that it should drop the condemnation of the Calvinists, since he regarded himself as a bridge builder between the confessions. He wrote to this effect to Duke Johann Friedrich of Wurttemberg in the spring of 1609. In the end, this exchange of correspondence on the subject went on until 1625.

Since Kepler refused the ubiquity of the person of Christ, his attempts to obtain re-employment in Wurttemberg, which he had requested from the Duke and the Duke's widowed mother Sibylla, were brought to the attention of the consistory. They issued a conclusion with an absolutely negative result, insinuating that Kepler was a Calvinist in disguise. Since this was to no avail, Kepler ended up staying in Linz.

Kepler's approach to the doctrines in question was a rational one, based on his scientific understanding of reality. He thus rejected the omnipresence of Christ, just as Calvin had done, and therefore also his local presence during the celebration of the Last Supper. Kepler argued about Christ's spatio-temporal nature, and the idea that he resided in heaven since ascension. God had created all locations and space as a whole. He needed no preferred place on Earth but was present everywhere through his actions. But since Christ himself resided in heaven he had to be absent from any earthly place. This was his stance, when Hitzler excluded him from participating in the Last Supper ceremony, the sacrament, in Linz.

And more trouble was initiated from another quarter. The governor Hanns Jakob Loehl, acting on behalf of Rudolf II, ordered the suspension of the school Kepler was employed at in 1600 during the course of the Counter-Reformation. However, the school continued to exist clandestinely until 1609, when it re-opened again officially after an imperial edict of restitution.

Meanwhile, Kepler's conflict with Hitzler dragged on, and he discussed the reasons why he would not sign the formula of concord with other clergy, who were prepared to let him participate in the Lord's Supper. Hitzler himself was confirmed in his resolution by the Wurttemberg consistory, who argued that Kepler would deny the omnipresence of Christ, and therefore sided with the Calvinists on this point. So the impasse continued. In addition, Kepler befriended Erasmus von Starhemberg, who had Calvinist leanings and who lived in Eferding, where his second wife, Susanne Reuttinger, came from.

In spite of all these difficulties, Kepler took his responsibilities as a Christian paterfamilias seriously enough to write a small instruction book for the members of his family, based on his own interpretation of the Holy Sacrament, which refers to the body and blood of Christ. This was printed in Prague in 1617. As sources, Kepler referred to various liturgical agendas, in fact three altogether: the official Wurttemberg version, a version revised by Hitzler for Linz, and the traditional one by the German theologian Veith Dietrich from Nuremberg. In the end, Kepler's consolidation presented a kind of mixture, especially with regard to the composition of the words of institution.

Of course Kepler's writings came to the attention of the consistory in Wurttemberg. Hafenreffer and his collegues interpreted this small instruction book for their private use as a kind of manifest of Kepler's beliefs and at the same time a critique of the official order of the divine service. This was the starting point for the long exchange of letters between them, notably Hafenreffer and Maestlin, and Kepler. But in Linz Kepler was also regarded as a heretic because of his confessional attitudes. Kepler defended himself by asserting that his interpretation of the scriptures could not be regarded as anything of an innovation. In his view, the positions of the Lutherans and the Calvinists were closer than their mutual condemnations would suggest.

Both Maestlin and Hafenreffer tried to moderate Kepler's temper and position, since they feared he could be in personal as well as spiritual danger. In those days, exclusion from the church meant exclusion from society as well.

During the course of the witch trial against his mother (see later in this chapter), Kepler returned to Wurttemberg for a short visit and met Hafenreffer in person to ask him to intervene on his behalf with Hitzler, since he hoped to be able to take part in the Lord's Supper, but in the end Hafenreffer refused. The further discussion in their exchange of letters reads like a dialogue of the deaf. Hafenreffer argued as a Lutheran theologian, whereas Kepler used the rationality of a mathematician and astronomer to tackle the question of the omnipresence of Christ. Hafenreffer sensed this attitude and at one point he remarked: "Mathematician—you are beginning to become stupid." He insinuated that, by using geometrical constructs to discuss theological truths,

Kepler would muddle his mind. In response to this, Kepler argued that Hafenreffer's position and the position of other theologians could only be believed when he shut his eyes as a geometer. Johannes Kepler was thus a perfect representative of the tragedy of the confessional era between the Middle Ages and modern times. Hafenreffer's last letter to Kepler on this matter dated from 31st July 1619.

Details about the above the question of Kepler's credo can be found in J. Huebner, "Die Theologie Johannes Keplers zwischen Orthodoxie und Naturwissenschaft", Tuebingen, 1975 [24].

Credo

In 1618, Kepler drafted another theological work, which was published anonymously in 1623. It was called: "Glaubensbekandtnus und Ableitung allerhand desthalben entstandener ungütlichen Nachreden." Approximately translated, this becomes "Confession of Faith and Deduction of Defamations Caused by it." A hundred copies were published by Bernegger in Strasbourg. Basically, it was a summary of Kepler's quarrels with church authorities and the defense of his own position.

In short, he based his own beliefs on patristic teachings and accused the three confessions Catholicism, Lutheranism, and Calvinism of having torn apart the truth, which he, Kepler, now had to put back together again piecemeal. There was never any reaction to this tract from Wurttemberg. When Bernegger tried to assist Kepler in his application for a position in Strasbourg by showing it to one of the councilors, the latter objected to any possible appointment on confessional grounds.

There is one remarkable detail in the tract which deserves special mention. In his argumentation, Kepler referred in some detail to the book "De republica ecclesiastica" by Marcus Antonius de Dominis.

Later, in 1624, De Dominis died as a heretic in a dungeon in Rome, but for Kepler he had been a potential savior of Christianity. In his book "De stella nova in pede serpentarii, et qui sub ejus exortum de novo iniit, trigono igneo" (On the Meaning of the Nova of 1604 at the Foot of the Serpent Bearer), which was published in 1606, he speculated about the possibility that a man would one day arise and solve the conflict between the confessions by convening a council to that end. Kepler addressed this topic again in a letter to Herwart von Hohenburg, a Bavarian statesman and scholar, in which he compared the celestial spheres to the hierarchy of the church. Treating the discovery of the nova with the analytical instruments of his prognostica, he saw his vision fulfilled in the person of de Dominis. Kepler pronounced himself in a similar way in another letter to Hafenreffer on the 28th of November 1618, in which he claimed that de Dominis would be the right man to heal the wounds of the religious war which he, Kepler, saw coming. Of course, after the death of de Dominis, Kepler's hopes and the validity of his prognosticum melted away to nothing.

Johannes Kepler, ensnarled in religious conflicts and quarrels and having devoted his scientific attention to other things, finally tried to return to astronomy in 1615, but had to abandon the idea of continuous work on this matter because his attention was drawn to some dramatic events surrounding the fate of his mother.

Witch Hunt

Kepler's sister Margarethe, who was married to a priest named Georg Binder in Heumaden near Stuttgart, informed him that his 70 year old mother Katharina Kepler had been charged on an account of witchcraft. She was in danger of being burned at the stake. The court proceedings would drag on for six years from 1615, when she was accused for the first time, to 1621, when she was released on September 28th. This was against a backdrop in which a significant number of so-called witches were being executed in the vicinity of his mother's place of residence: 6 in Leonberg and 38 in Weil der Stadt between 1615 and 1629. Among the victims was the foster mother of Kepler's mother, Renate Streicher.

Kepler's personal intervention avoided the worst for his mother, including torture. Twice he had to travel from Linz at his own expense, since his personal appearance was required from October to December 1617 and again from September to October 1621.

Kepler's mother satisfied most of the criteria that people associated with witches regarding her external appearance: she was of dark complexion, lean, and small. As already mentioned, she was a cantankerous individual. She was known to be a naturopath. At the beginning of her trial, she was accused of all sorts of misfortunes which had taken place in Leonberg in recent history: cattle disease, the death of a family father, the paralysis of the village tailor, and other such events. There were witnesses claiming that she could walk through closed doors and that she had committed blasphemy by denying the resurrection.

The incident which triggered her trial was the abdominal pains of the wife of a glazier in Leonberg, Ursula Reinbold, who happened to be the sister of the barber of Prince Archilles of Wurttemberg. This barber was unable to cure the disease. His explanation was that it must have been the result of sorcery and could only be eliminated by the person who had bewitched the poor woman in the first place. So they went and asked Kepler's mother, who had a reputation as a healer and had administered a healing brew to the woman some two years earlier. The ulterior motive was of course to convict her as a witch in case of any medical success. They also remembered her relationship to the convicted witch Renate Streicher. But Katharina Kepler's other son Christoph reacted by filing a libel suit, which, however, was turned down, and the judge responsible now began to support the glazier's wife in her efforts to strive for a witch trial against the accused. The Reinbold's suit against Katharina Kepler asked for 1000 Guilders in compensation for the pain suffered by Ursula. To support their claims, they brought forward evidence of damages by sorcery suffered by other persons, who they claimed had died by pure physical contact with Katharina. Other victims included a paralyzed schoolmaster, a butcher who suffered unknown pains, a dead pig, and a mad cow. Altogether 49 articles were filed against her and up to forty witnesses had been called by the end of the proceedings.

It was then that Johannes Kepler intervened with all his authority as Imperial Astronomer. In the first place, he asked his mother to come to Linz, where she stayed for about nine months, before returning to Leonberg with him. He then requested all relevant documents from the trial, and at that point the trial was suspended. However, the search for new evidence against her continued, and Kepler continued to refute it. The process dragged on until 1520, when Katharina Kepler was suddenly arrested with the intention of finishing the case through the use of an accelerated procedure. A date for torture had been fixed, and this prompted her son to hurry to Leonberg and intervene. Kepler succeeded in arranging for all the case records to be forwarded to the faculty of law at the University of Tubingen for assessment. One member of the faculty was Christoph Besold, a personal friend of Kepler. The faculty recommended showing the torture instruments to the accused. He himself had drafted a defense of 128 pages. If she remained steadfast in her denials, then she should be released, but carry the costs of the whole procedure. Katharina Kepler died on April 13th 1622, half a year after her release.

Harmonices Mundi

Finally, in 1619 Kepler completed his magnum opus which he had been working on with interruptions while editing and compiling the Tabulae Rudolphinae. Its title was:

"Joannis Kepleri Harmonices Mundi Libri Quinque" (The Five Books of Johannes Kepler's Harmony of the World) (Fig. 7.3).

He dedicated the work to King James I of Great Britain.

Large parts of "The Harmony of the World" give the impression of a textbook on music theory, especially the third book, the "real harmonic book", as Kepler called it. In it he worked out a complete theory of music, extensively taking into account the work on music theory by Vincenzo Galilei (1520–1591), the father of Galileo. "Music", according to Kepler, would help him to "discover the blueprint of creation" [25].

With respect to the history of the genesis of his "Harmony of the World" Kepler writes in 1619:

> When some twenty-four years ago I started making these observations, I first examined whether the planetary spheres were separated by the same magnitude from each other. In the end, I arrived at the five spatial figures. With this I obtained a number of planetary bodies and magnitudes of the distances which were nearly correct [...].
>
> Astronomy had been perfected during the last twenty years; but nevertheless, the distances still did not match with the spatial figures, and at the same time they did not show any discrepancies concerning the eccentricities distributed in such an uneven way across the planets [...].

Ioannis Keppleri

HARMONICES
MVNDI

LIBRI V. QVORVM

Primus GEOMETRICVS, De Figurarum Regularium, quæ Proportiones Harmonicas conſtituunt, ortu & demonſtrationibus.
Secundus ARCHITECTONICVS, ſeu ex GEOMETRIA FIGVRATA, De Figurarum Regularium Congruentia in plano vel ſolido:
Tertius propriè HARMONICVS, De Proportionum Harmonicarum ortu ex Figuris; deque Naturâ & Differentiis rerum ad cantum pertinentium, contra Veteres:
Quartus METAPHYSICVS, PSYCHOLOGICVS & ASTROLOGICVS, De Harmoniarum mentali Eſſentiâ earumque generibus in Mundo; præſertim de Harmonia radiorum, ex corporibus cœleſtibus in Terram deſcendentibus, eiuſque effectu in Natura ſeu Anima ſublunari & Humana:
Quintus ASTRONOMICVS & METAPHYSICVS, De Harmoniis abſolutiſſimis motuum cœleſtium, ortuque Eccentricitatum ex proportionibus Harmonicis.
Appendix habet comparationem huius Operis cum Harmonices Cl. Ptolemæi libro III. cumque Roberti de Fluctibus, dicti Flud. Medici Oxonienſis ſpeculationibus Harmonicis, operi de Macrocoſmo & Microcoſmo inſertis.

Cum S. C. Mtis. Priuilegio ad annos XV.

Lincii Auſtriæ,
Sumptibus GODOFREDI TAMPACHII Bibl. Francof.
Excudebat IOANNES PLANCVS.

ANNO M. DC. XIX.

Fig. 7.3 Title page of Harmonices Mundi

Thus by and by, especially during the last three years, I arrived at the harmony by tolerating small deviations from the spatial figures. I was led to this, on the one hand, by the thought that the harmonies would play the role of the forms, which would apply the final touches, while the figures would play the role of matter, which is represented in the world by the number of the planetary bodies and the raw extent of the spatial realm. On the other hand, the harmonies deliver the eccentricities as well, which the spatial figures cannot even promise. Hence, the harmonies gave nose, eyes, and the remaining limbs to the statue, while the spatial figures prescribed only the external size of the raw matter.

In this way, the truth of the Pythagorean idea of celestial harmonies became the guiding principle of his research. So he tried to support his own discoveries by a re-interpretation of the Pythagorean concept of the music of the spheres. His planetary laws almost became a mere by-product of his investigations into this music of the spheres.

Pythagoras had himself come to the conclusion that there must be some sort of connection between mathematics and music. He derived this from the fact that there was a relationship between the pitch and the length of the string on a musical instrument, for example. Harmonic intervals could be represented by proportions of whole numbers. In this way, sound was nothing else but embodied numbers. Pythagoras transferred this finding to the order of the cosmos as represented by the regular motions of celestial bodies, thinking of some kind of harmony of the spheres on which those bodies were moving. Everything could thus be represented by numbers.

The order of the world was the result of its mathematical structures. Just as music had a mathematical core, the rest of nature must be based on mathematical laws.

Thereafter, the Pythagoreans' theory of music was lost for some time, but then rescued from oblivion by Ancius Manlius Severinus Boethius, who lived from 480 to 524 AD. In the five books of his "De institutione musica", he divided music into a hierarchical structure:

- Musica mundana—the music of the spheres
- Musica humana—the interplay between body and soul, and also between humans
- Musica instrumentalis—voices and instruments.

Thus Kepler writes in his Harmony of the World:

> Therefore celestial motion is nothing but continuous part music (to be captured by the mind not the ear), aiming at certain previously designed six-part conditions (as though in six voices), caused by dissonant stress and somehow through syncopations and cadences (just as man applies them by mimicry of these natural dissonances) and thus defining different attributes through the immeasurable passage of time. It is therefore no surprise that man, the imitator of his creator, has finally discovered the art of vocals involving several voices, unknown in the past.

According to Kepler, real musical harmonies are nothing but material realizations of the harmonies of the spheres in exactly the sense of Pythagoras and Boethius. Kepler believed in the truth of this and tried to prove it by studying the motions of the planets. In his imagination, he was convinced this would provide him with an insight into the thoughts of God.

At this time, Kepler was by no means the only person trying to harmonize the cosmos by resorting to musical elements. The English physician Robert Fludd, who lived from 1574 to 1637, published an esoteric book entitled "Ultriusque cosmic maioris scilicet et minoris Metaphysica, physica atque technica Historia", in which he related the macrocosm, i.e., the Universe, to the microcosm, the world of man. In this book, he proposed a "cosmic monochord" as the basic element for his world harmony. This approach was refuted by Kepler as unsuitable. Kepler argued that Fludd was mixing up physical quantities of different dimensions and ignoring empirical facts.

Kepler speculated that musical sense and the preference for harmony might be the product of evolution, and harmonic proportions might be "born into humans". In any case, he believed that God had observed two important principles when he created all things: a geometrical one, which gives importance to the spherical shape, and a harmonic one, which triggers the music of the spheres.

Concerning a possible harmony between the six planets known at that time, he speculates: "If there had been a single six-fold harmony or one particular distinguished one among several, one would be able to discover the constellation of the world at the point of its creation." This sounds pretty much like a comment in our time on the discovery of the cosmic background radiation.

Of course, it was not easy for Kepler to follow the power of persuasion of Brahe's observed data and accept ellipses for the planetary orbits instead of circles, which according to him and his contemporaries were the most perfect geometrical figures. But, as he wrote: "When the calculated values do not correspond, all our work will have been in vain." Thus the correctness of a theory is decided by its comparison to observation, to experiment, according to Kepler, even when experimental data are contrary to authoritative opinions or thousand-year-old dogmas.

And he writes: "Concerning the opinions of the saints about the natural things, I respond with this single word: in theology, the weight of authority applies, but in philosophy, that of rational grounds." This way of thinking and this attitude makes Kepler one of the most important founders of modern natural science. Let us now have a look at the structure of the work itself.

The first book develops the meaning of "harmony" from geometrical considerations. It is founded on Euclid's "Elements". Kepler tries to work out which of the regular polygons can be constructed with the help of compasses and ruler, i.e., using circles and lines, and which cannot.

In the second book, he investigates the congruency of regular polygons. However, his use of the term "congruency" is different to ours today. In his understanding, congruency says something about the ability of a regular polygon to completely or partly fill in a plane surface with equal polygons or to be able to form regular or partly regular bodies.

The remaining three books deal with music and the relationship between astrology and astronomy, together with other mystical concepts.

Kepler was looking for the musical intervals, those already known to Pythagoras, between the six planetary orbits. Once again, he had to try out a multitude of permutations. In the end, he found two quantities satisfying his expectations: the maximum and minimum orbital speeds of the planets, i.e., the speed at the perihelion of the ellipse, nearest to the Sun, and the speed at the aphelion, furthest from the Sun. Expressing the ratio of one to the other, this resulted in the following values:

- Saturn: 5:4 or major third
- Jupiter: 6:5 or minor third
- Mars: 3:2 or fifth
- Earth: 16:15 or half-tone

- Venus: 25:24 or diesis
- Mercury: 12:5 or beyond the octave.

Concerning the audibility of the music of the spheres, Kepler explained that it was limited in the sense that it could only be captured by the mind and not by the ear. He thus gave in completely to his requirement of harmony as the driving force behind his research, taking it to its very limit, until he had lost any ground beneath his feet. Those who succeeded him would have to climb down from these heights until they reached the almost complete dissolution of this harmony. Before we turn to Kepler's three laws, here is a short diversion quoted from the text:

> Of course, there are no notes in the sky and the motion is so fast that, by the friction with the celestial air, some sort of buzzing or whistling should occur. This would leave very little light. If this communicated anything about the orbits of the planets, it would either communicate it to the eyes or to a similar sensory organ.

The fifth book contains Kepler's third law in Chap. 3, which he had discovered shortly before the famous defenestration in Prague in 1618 which triggered the Thirty Year War. It is remarkable that the three laws were used as a basis to justify his mystical assumptions.

Against all expectations, Kepler's "Harmony of the World" was not listed on the Index of Forbidden Books by the Holy Officium, while his text book "Epitome astronomiae Copernicanae" (Layout of Copernican Astronomy) was.

Kepler's Three Laws

Here are Kepler's three laws, which are still the cornerstones of planetary astronomy and presented in any serious textbook to this day:

I. The planets move in elliptical orbits with the Sun situated at one of the focal points.
II. The radius vector from the Sun to a planet sweeps out equal areas in equal times.
III. The square of the period of any planet T is proportional to the cube of the semi-major axis a of its orbit:

$$T_1^2/a_1^3 = T_2^2/a_2^3 = T_3^2/a_3^3 = \dots$$

Kepler's laws are valid for all periodically recurrent celestial bodies in our Solar System, i.e., for the planets, but also for the Earth and its moon, and of course for artificial Earth satellites. Kepler's observation used to derive his spherical harmony based on the maximum and minimum orbital speeds at the perihelion and the aphelion is a consequence of his second law. It agrees with the law of conservation of energy, which states that the total energy of an orbiting body remains constant: the potential energy is at its minimum at the perihelion and at its maximum at the aphelion, while the kinetic energy, and thus the speed, is at its maximum at the perihelion and at its minimum at the aphelion.

With certain simplifications, it is possible to derive the law of gravitation from Kepler's third law. This then leads directly to the actual cause of the observed motions.

Confusion and Continued Creativity

In 1616 Francesco Ingoli, a future member of the Index Congregation who played a prominent role in the proceedings against Galileo, published a pamphlet against the Copernican system, entitled "De situ et quiete Terrae contra Copernici systema disputatio". Kepler was asked by the intermediary Tommaso Mingoni to reply to this tract. It took Kepler two years to respond, but he thoroughly demolished Ingoli's weak arguments. To strengthen his own reasoning, Kepler had attached a copy of his "Epitome" to the correspondence. But Ingoli responded that he did not feel intimidated, even by such an authority as Kepler. Galileo was well aware of the whole dispute.

Shortly after the onset of the Thirty Year War, the Emperor Matthias II died, and his cousin was crowned in Frankfurt to become Emperor Ferdinand II. He allied himself with the Bavarian Duke Maximilian, effectively pawning Upper Austria to win his support for his war in Bohemia. So in July 1620, Maximilian moved to Linz, and war finally reached this provincial town. After Ferdinand had decided the Battle of the White Mountain in his favor, he began the ruthless eradication of Protestantism in Bohemia. But Johannes Kepler once again remained an exception. He continued in his post as Imperial Mathematician under Ferdinand.

As already mentioned, while working on his "Harmony", Kepler worked in parallel on a textbook called "Epitome astronomiae Copernicanae". This book was intended to popularize and explain the Copernican world model. In it, Kepler inserted elements of his own thinking, which appeared in full in "Harmonices Mundi". In typical textbook manner, the work consisted of a series of questions and answers. Kepler started on it in 1615 and finished it in 1621. The "Epitome" was divided into three volumes which were published successively in 1617, 1620, and 1621. Each volume contained in turn separate books or chapters—altogether seven. Among others the "Epitome" also contained Kepler's third law. Volume I was put on the Index of Prohibited Books on February 28th 1619.

Ever since his first encounter with Tycho Brahe, Kepler had been occupied with the edition of the Tabulae Rudolphinae. He had been working on and off on these astronomical tables ever since. The mathematical calculations were tedious and time-consuming. Then, in 1617, just when he was beginning his Harmony of the World, a booklet found its way into his hands which described a method for simplifying his task. It was called "Mirifici Logarithmorum Canonis Descripto" (Description of the Wonderful Canon of Logarithms) by the Scotsman John Napier, Lord of Merchiston, a mathematician and theologian. Napier had derived the word "logarithm" from "logarithmanteia", meaning something like "prediction of word by number and vice versa", coined first by the German mathematician and astronomer Caspar Peucer in his "Commentarius des praecipius divinationem generibus" (Commentary about Kinds of Predictions), published in 1553. Logarithm s are mathematical tools that facilitate complex calculations. They are representations of real numbers in powers relative to the same base, usually base 10. Napier presented logarithmic tables in his booklet, but since he did not explain how he had derived them, Kepler thought them to be untrustworthy and created his own tables by resorting to procedures from the 5th Book of the Elements by Euclid. He dedicated the book, in which he described his methods in detail, to Philipp III, landgrave of Hesse-Butzbach, who had it printed in Marburg in 1624.

After 22 years, Johannes Kepler finally fulfilled his contract with the Empire by completing his work on the Tabulae Rodulphinae. This was in 1624. Although the contents and substance were accomplished, finalization for printing dragged on due to quarrels with Brahe's heirs, lack of money, and the search for a suitable printing house. In fact the printing house of Johannes Plank in Linz, which Kepler favoured, was burned down during a peasant uprising, while at the same time soldiers took quarters in Kepler's house, which made continuous work impossible. The two events finally prompted Kepler to leave Linz together with his family.

Chapter 8
Between Ulm and Prague

1626–1628

Ulm is situated on the Danube at the frontier between Baden-Wurttemberg and Bavaria, at the southern edge of the Swabian Alps, some 170 km north of Munich, and today counts more than 125,000 inhabitants. It is famous for its Gothic minster with the highest church spire in the world, reaching 161.53 m (Fig. 8.1). During the Late Middle Ages, it became a free imperial town because of its privileged situation at the crossroads of important trading routes. Around 1500, the town reached the pinnacle of its urban development. During the turmoil of the Reformation, Ulm became Protestant, but it lost its independence and was subordinated to Emperor Charles V. The town lost still more of its riches and prosperity during the Thirty Years War.

Tabulae Rudolphinae

Although Johannes Kepler had finished editing the Tabulae Rudolphinae, commenced by Tycho Brahe, after 22 years of work, he could not have them printed in Linz, as already mentioned in the last chapter. He relocated to Ulm and lived in Kohlgasse 8, the house of a physician. Here he completed the tables for printing in September 1627. They were published by Jonas Saur under the full title:

> Tabulae Rudolphinae, quibus astronomicae scientiae, temporum longinquitate collapsae restauratio continentur.

These tables were preceded as a reference work by the Alfonsinean Tables from the 13th century and the Tabulae Prutenicoe Coelestium Motuum, by Erasmus Reinhold, from 1551. Its most important compilation is a collection of tables and rules for the prediction of planetary positions. The mean deviation between calculated and observed planetary positions in the older Tables had been 5°, whereas in the Tabulae Rudolphinae it had been reduced to 10'. Besides containing a further catalogue of 1005 fixed stars with their positions, it comprised refraction tables, logarithms, and a list of important towns around the world.

© Springer Nature Switzerland AG 2020

W. Osterhage, *Johannes Kepler*, Springer Biographies,
https://doi.org/10.1007/978-3-030-46858-3_8

Fig. 8.1 The tower of Ulm cathedral http://de.wikipedia.org/wiki/Bild:Ulm_muenster.JPG

Altogether the book contained 120 pages of text and another 119 pages filled with the tables themselves. The frontispiece was designed by Kepler himself and executed by Georg Celer, an engraver from Nuremberg. It shows important astronomers gathered in the temple of Urania (Fig. 8.2). From a commercial point of view, the book was no success. In 1628, to increase its attractiveness, Kepler created a supplement called "Sportula" (4 pages) with instructions on how to use the tables also for astrological purposes.

The original manuscript had been safely carried by horse and cart along the Danube from Linz to Ulm, and not by boat since the river was iced over at the time. It was the turn of the year 1626/1627. The printing process was anything but smooth. There were constant quarrels between Kepler and the printing master. And quarrels continued in parallel with Brahe's heirs about the purchasing price of the book, which was finally fixed at 3 Guilders at the Frankfurt book fair in October 1627, under the mediation

Fig. 8.2 Frontispiece of the "Tabulae Rudolphinae"

of a commissioner. Since Tengnagel had died in 1622, Brahe's son Georg wrote the introduction to the work, which had always been assigned to Brahe's heirs.

Another acquaintance of Johannes Kepler lived in Ulm. This was Johann Baptist Hebenstreit. He was the schoolmaster at the grammar school there. He had been involved in the famous Comet Controversy in Ulm, in which a number of mathematicians, philosophers, and theologians had argued about the significance of the comets. These had been discovered by Kepler, among others, at the onset of the Thirty Years War in 1618. The controversy was about whether these were just natural phenomena or signs of the wrath of God and a warning of retribution. The controversy ended in a draw and was closed under the mediation of René Descartes in 1619.

However, when Hebenstreit learned that Kepler was on his way to Ulm, he induced the City Council to ask Kepler to bring some order into the weights and measures in use in Ulm at the time. Kepler complied with a system based on the form of a brass basin which contained all gauged measures, i.e., length, volume, and weight. This basin can still be seen in the museum in Ulm. Hebenstreit later composed a long Latin poem about the meaning of the frontispiece of the Tabulae—altogether 462 verses, which was attached as an epigram to the main body of the compilation.

Return to Prague

After completion of the Tabulae, there was no more business for Kepler in Ulm, so he relocated to Prague for a short visit at Easter 1628. At the time, festivities for the coronation of Emperor Ferdinand's son as King of Bohemia were in full swing. Kepler was received by the Emperor and handed a copy of the Tabulae to him, after which he received the promised 4000 Guilders in return for a letter of dedication accompanying the book. The money was to be provided fifty-fifty by the cities of Ulm and Nuremberg. The remuneration was initially conditioned to Kepler's conversion to the Catholic faith, but he refused. He continued his refusal even when offered the post of professor of mathematics at Prague University. His argument was that, in his heart, he already was effectively a Catholic, since he belonged to the community of all baptized Christians, although that had nothing to do with any official adherence to the Roman Catholic Church. That much he had already stated in his correspondence on mathematical problems with the Jesuit Paul Guldin, an astronomer and mathematician in Graz and Vienna.

Wallenstein

Hermanitz is a small village on the river Elbe in eastern Bohemia. From 1548 to 1623, it belonged to the dynasty of von Waldstein. The church, dedicated to Mary Magdalene was rebuilt several times during its centuries-long history. On either side of the altar, there are almost life-size funerary monuments depicting a knight and his spouse. The Czech inscriptions around the relief identify the knight as Vilim the Elder of Waldstein of Hermanitz, who died in 1595, and his wife Markyta of Smilice, who died in 1593. The monument was erected under the authority of their only surviving son, Albrecht Wenzel Eusebius, after his return from an educational journey to France and Italy at the turn of the sixteenth century. In Italy he visited Padua, where Galileo was teaching astronomy at the university in 1601, but it is uncertain whether they met each other. Waldstein—or Wallenstein as the man was known later—was accompanied by the German mathematician Paul Virdung. This is documented in a letter Virdung wrote to Kepler in 1603, in which he recounted his trip with Wallenstein.

The German historian and scholar, Golo Mann, son of the Nobel laureate Thomas Mann, traced the name and origin of Wallenstein back to the 12th century [26]. According to his research, the original name of the later general was Waldnstein, but this was difficult for the Bohemians to pronounce. So they called his ancestors Waldstein, whereas the native German speakers changed it later to Wallenstein or von Wallenstein. But again, this was the result of a name change. As is not uncommon in many countries, nobility and other gentry derived their family names from the estates they were living in. This was the case for Wallenstein as well. The castle of Waldstein—translated as wood rock or forest rock—was built in a location surrounded by woods near Turnau in what is today the Czech Republic by a man who was called Zdenek before he named himself after his newly constructed castle and thus genealogically became the ancestor of Albrecht Wallenstein, the person of interest in this chapter (Fig. 8.3).

The first (indirect) encounter between Wallenstein and Johannes Kepler took place in the year 1608, while Kepler was first in residence in Prague. Wallenstein contacted the astronomer—although it would be better to say astrologer in this case—via the intermediary of Dr. Stromair, a physician. However, Stromair did not mention the name of the person on whose behalf he was acting, only that he was a member of the nobility. Stromair gave the date and the hour of birth: 24th September 1583 of the Gregorian calendar at 04:30 a.m., since his business was to ask Kepler to issue a horoscope for his client. Kepler complied with this request, which for him was nothing unusual, and handed his results over in the summer. Kepler specified the hour of birth at 04:30 a.m. and 1 ½ min.

Fig. 8.3 Wallenstein: Duke of Friedland, member of the Imperial Council of War and Chamberlain, Highest Colonel of Prague and General alike; copperplate engraving 1625/28

Kepler described the unknown client as someone with "an alert, enthusiastic, assiduous, and restless mind, curious and keen on all kinds of innovation, a person who does not like meanness and trouble-making, but rather prefers to look for new or untested means that may still seem strange, but otherwise keeps many more thoughts for himself than he would allow to be glimpsed or otherwise sensed from the outside."

In short, the horoscope characterized Wallenstein as an ambitious individual who would attract all kinds of enemies during his lifetime. Kepler forecast a marriage with a wealthy woman at the age of 33, while in fact this took place seven years earlier. For his later years, Kepler predicted more displeasing developments, and among others that Wallenstein would become the leader of a discontented mob, but he could also promise him an awe-inspiring and dangerous life.

This was the first version of Kepler's horoscope for Wallenstein. But Wallenstein was convinced that there was a strong relation between the relevant constellation precisely at the moment of birth and the further fate of the person in question—contrary to Kepler, who thought that a constellation of stars would only provide some sort of impulse for the further development of a person [27]. He thus compared his own fate to the predictions of the horoscope over the next sixteen years. Some of these predictions were fulfilled, others were not, some at an earlier date than expected, others at a later date. So on several occasions, he requested a review of the horoscope with reference to a modified hour of birth. In the end, Kepler calculated five versions of the horoscope corresponding to a time span of altogether 40 min. His rectified version of 21st January 1625 was based on 04:00 36 ½ a.m. as the time of birth. But Kepler saw no reason to modify his conclusions from the 1608 version. Wallenstein was not satisfied with this and asked for a detailed elaboration at the end of September 1625. In particular, he wanted to know whether his death would be caused by a stroke and this outside his home country? Also whether he would receive honors and wealth outside his home country? And whether he would be able to continue in the art of war, and in which country, and with how much luck? He wanted to know who his enemies would be and whether it was true that his worst enemies would be the Bohemians, his fellow countrymen.

Until 1625, Kepler was unaware who his client was, but suddenly one of the intermediaries, Lieutenant Gerhard von Taxis, revealed the secret by citing Wallenstein's name and position in a letter transported by a young officer from Vienna, Christoph von Hochkircher. Kepler responded that, now that he knew the true identity of his client, he had not received the agreed fees. He had adjusted the hour of birth as requested, but his answers to Wallenstein's detailed questions were of a very general nature. There was nothing really new in this version of the horoscope, except some minor adjustments regarding the years, for which he had predicted some changes in the fate of his customer. So in the end there were two famous versions of Kepler's horoscopes for Albrecht Wallenstein: those of 1608 and 1625. Copies of them exist in various renowned libraries. As to the question of remuneration, the cadet Hochkircher handed him some devalued twenty shilling pieces, claiming that it was too dangerous to carry the bulk of the money on the road. When Kepler arranged for an advocate named Miller to collect the full sum, Hochkirchen had already spent the rest of the sum for his own purposes.

Later, Wallenstein insisted that he needed to consult his horoscope again before meeting the King of Hungary in Regensburg. But that was after Kepler had already accepted an employment in Sagan, offered to him by Wallenstein. The background was that Kepler was still trying to obtain the 11,817 Guilders the imperial court owed him for his many years of service and which had not been paid until he met Wallenstein in Prague. Wallenstein proposed a different solution, based on contributions from the imperial bourse, over which he had control at the time. Kepler was promised 1000 Rhenish Guilders per year and a weekly allowance of 20 Guilders for printing costs. So, in full view of the world, Wallenstein appointed Johannes Kepler as court astrologist in Sagan, a town in lower Silesia. Kepler accepted the deal and relocated to Sagan with wife and child.

Astrology

Johannes Kepler produced a substantial collection of astrological works. Among them were the Prognostica, almanacs, and horoscopes, but also several treatises about astrological questions. To the latter belong "De fundamentis astrologiae certioribus" ("About the Secured Foundations of Astrology"), dated 1609, and "Tertius interveniens", a somewhat longer essay about the subject from 1610. In his prognosticum for 1618, the year the Thirty Year War started, he prophesized: "It is sure that the month of May will not pass without major difficulties in certain places and with certain quarrels which are currently in the making." Kepler's legacy included horoscopes for more than nine hundred people, including personalities from society, contemporary rulers, and members of his family. The earliest ones date back to 1592, shortly after his master's degree in Tubingen. His horoscopes comprised a figure and a written interpretation. The figure consisted of a number of squares, the square in the centre containing the name and the date of birth of the person. Other information concerned the astronomical position of the astrological houses, the planets, and the moon. The whole was than analyzed and a written statement offered as an interpretation of the astronomical data.

For us today it seems strange that astrology was once regarded as a science. As recently as 2009, the German information scientist Walter Oberschelp tried to formulate a rational assessment of this matter in a renowned computer journal [3]. For him, everything centered on the basic question that might be asked by someone looking at the sky above: how can the sky above us and the processes going on there be understood in time and space? In the end, this approach led to quantitative evaluations. But in spite of that, there was a persistent desire to find relationships, preferably simple ones, and use them to predict the future. The old desire to detect commensurabilities has been widespread until modern times. Calculating the future meant calculating the positions of the stars, and recognizing the intentions of the gods. This is still present in today's esoteric astrology.

In spite of all the reservations of scientific astronomy with regard to the interpretive methods of astrology, the motivation for astrology has to be taken seriously. The most important indicators are the so-called “aspect s” of the horoscope, which are whole number angular ratios relative to the full circle, i.e., angular commensurabilities of star positions. Conjunctions and oppositions of celestial bodies, i.e., equal or opposite apparent angles on the ecliptic, are thought of as being especially meaningful, as are angles of 90°, 60°, or 120°. This is the age-old fascination with whole number ratios. As we know, the cosmos contains many completely exact and enigmatic harmonies: it is nearly a rule that not only our own moon, but also the natural satellites of many other planets in close orbits move in 1:1 synchronous rotation. This is true for all the larger moons of Jupiter. Even the ratios of their orbital periods can be expressed in part by whole numbers. Mercury possesses a 2:3 synchronicity between its rotational and orbital periods. The exact ratio of 2:3 between the orbital periods of Neptune and Pluto is also remarkable.

Chapter 9
Sagan and Death in Regensburg

1628–1630

The final version of Kepler's horsocope for his new master contained a serious and very specific warning for the beginning of the year 1634. It was set to occur in or near a castle in the extreme west of what is today the Czech Republic.

Eger and Its Castle

The ancient Bohemian town of Eger (Fig. 9.1) is called Cheb today, with its 32,000 inhabitants. It lies in the west of the Karlovarsky kraj or Carlsbad region on the banks of the river with the same name. During the Hussite war, the town participated in the anti-Hussite coalition. In later years, Eger joined the Lutheran Reformation, but Emperor Rudolf II organized the Counter-Reformation in 1626 to bring Eger back into line.

Eger is crowned by its castle.It was built at the beginning of the 12th century by the Hohenstaufen dynasty, from which a number of Roman-German kings and emperors emerged. Today the remains are still visible, with the famous Black Tower.

It was in Eger that Wallenstein's fate was finally sealed.

Wallenstein's Death

On the 24th of January 1634 the King of Bohemia and Emperor Ferdinand II signed a deposition document to dismiss Albrecht von Wallenstein from his post of supreme commander of the imperial army. The order went out to all generals, officers, and ordinary soldiers to cease their loyalty to the deposed generalissimo and to refuse obedience to Wallenstein's closest confidants, Field Marshal Christian Ilow and General of the Cavalry Count Adam Erdman Trcka von Leipa, his brother-in-law. The document closed with an instruction, which meant nothing but a death sentence:

© Springer Nature Switzerland AG 2020

W. Osterhage, *Johannes Kepler*, Springer Biographies,
https://doi.org/10.1007/978-3-030-46858-3_9

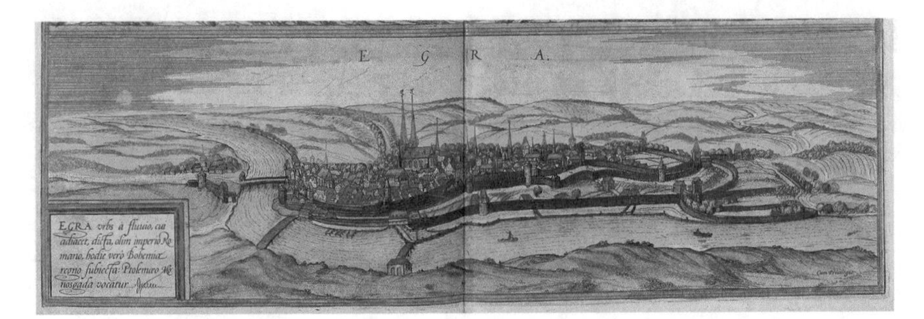

Fig. 9.1 Ancient view of Eger

"The head of the conspiracy and his major accomplices are to be taken prisoner, if possible, otherwise killed as convicted offenders."

In the face of these fateful events, Wallenstein had no choice but to flee and do everything he could to evade imperial authority. He chose to go further west to Eger, from where he hoped to make contact with the enemy operating in the Habsburg territories.

In the night of February 25th, three men of the island kingdom, John Gordon, a lieutenant colonel of Trcka's foot soldiers and at the same time commander of Eger, Walter Leslie, another lieutenant colonel of the same regiment, and the dragoon lieutenant Walter Butler decided to assassinate the fugitive and his followers.

Gordon invited Ilow, Trcka von Leipa, the latter's brother-in-law Wilhelm Kinsky von Vchynitz, and the cavalry captain Niemann, secretary to Wallenstein, to his quarters in the Castle of Eger for dinner, where Butler's men subsequently manned all entrances and exits. Between seven and eight o'clock, Butler's dragoons stormed the dining hall and finished off all the guests. Some minutes later, cavalry captain Deveroux ended the dramatic life of Albrecht von Wallenstein, once supreme commander of the Habsburg army, Count of Friedland, Mecklenburg, Sagan, and Glogau, by a single death blow in Pachelbel House near the market place, the house of the town commander, where his victim had elected to spend the night [28].

Sagan

Johannes Kepler did not live to witness the departure of his last employer, but during his stay in Sagan he had already had to bear the consequences of his master's demise.

Today Sagan or Zagan, a town in the rural district of Zagan in the Polish voivodeship Lebus, counts 26500 inhabitants. It is situated in Lower Silesia, halfway between the German town of Cottbus and the Polish town of Breslau. It came into the possession of Wallenstein in 1627, a year before Kepler arrived there for his final assignment. But first he went to Regensburg in May 1628, where his family had taken temporary residence. He made a detour via Linz to sell a copy of the Tabulae for 200 Guilders to his acquaintances there, on his way to Prague to hand in his resignation as imperial mathematician. Meanwhile, his family had arrived in Prague, and together they proceeded to Sagan, where Wallenstein had organized accommodation for them.

Of course, Wallenstein wanted a service in return: astrology. On the other hand, the prince was thinking of founding a university in Sagan, in which Kepler was to be the first cornerstone. Concerning this matter, Kepler wrote to his friend Matthias Bernegger that there might be an opening for Bernegger himself to obtain a chair there. But otherwise Kepler did not feel very happy in this small and remote place. He had no intellectual counterparts to converse with, and worse, the Counter-Reformation came to Sagan as well, and Wallenstein could do nothing against it.

Still, financially, Kepler's life was now somewhat easier. In 1630, he managed to get his ephemerides printed with forecasts for planetary positions on a day-to-day basis up to the year 1636. This became possible because he bought a printing press in Leipzig and had it transferred to Sagan. His assistant was a young man from nearby Lauban, Jakob Bartsch. Kepler became quite fond of Bartsch and finally arranged the marriage between him and his daughter Susanne, who lived in Durlach in Baden at that time. Kepler engaged the services of Bernegger once again to organize the wedding in Strasbourg on the 12th of March, and even stand in for him as the bride's father, since he himself could not leave his post in Sagan. According to a letter from Bernegger, everyone seemed to have been pleased with the arrangement. Pleased again was Kepler on the 18th of April of that year by the birth of his daughter Anna Maria. And while the printing press was working for the ephemerides, he even had leisure to work on a fantasy novel about a journey to the Moon called "Somnium seu opus posthumum Astronomia lunaris" ("Dream or Posthumous Work about the Astronomy of the Moon"), published by his son Ludwig in 1634. These three events were the last happy ones for Johannes Kepler in this fateful year 1630.

When Wallenstein was deposed, Kepler had once again to worry about his livelihood, and he remembered the outstanding debt the imperial court owed him: more than 10,000 Guilders. So he prepared a journey to Linz to settle things once and for all. His worries were aggravated by the fact that his own horoscope concerning his forthcoming sixtieth year of life forewarned of possible adversity since it showed astounding similarity with the year of his birth. After taking leave of his family, whose care he entrusted to his son-in-law Bartsch, he set off on his journey. The first stop was in Leipzig where he would stay for one week in the house of Professor Phillip Mueller, an old acquaintance. He took out a loan of 50 Guilders from Mueller and carried on to Nuremberg to contact another acquaintance, Phillip Eckebrecht, from whom he had requested a chart for the Tabulae Rudolphinae. The chart was not ready and Kepler moved on in the direction of Linz.

Fig. 9.2 Panoramic view of the southern Half of the Stone Bridge (Steinerne Brücke) in Regensburg; https://common.wikimedia.org/wiki/File:Regensburg_Steinerne_Brücke_Panorama_I.jpg

Regensburg

Today, Regensburg (Fig. 9.2) is the capital of the county of Upper Palatinate in Bavaria, and it now has about 150,000 inhabitants. It is at the same time the seat of the bishop of the Regensburg diocese. It is situated at the northernmost point of the Danube, at the confluence of the rivers Naab and Regen. During the Thirty Years War, Regensburg became a safe haven for Protestant fugitives from Austria.

The weather was foul. There was fog and continuous rain. Kepler was riding on horseback and arrived in Regensburg at the "Steinerne Brücke" (Stone Bridge) (Fig. 9.2), where he found accommodation in the house of a tradesman, Hillebrand Billi. Three days later he was struck by high fever. Emperor Ferdinand, who was in Regensburg at that time for the coronation of his 2nd wife Eleonore of Gonzaga, got news of the poor health of his former protégé and sent him his regards, together with a donation of 30 Guilders. The fever got worse and Johannes Kepler died on the 15th of November 1530, accompanied by a preacher. At that time, due to the presence of the emperor, many of Kepler's old friends and acquaintances were still in town, and they attended his funeral at the Protestant cemetery of St. Peter outside the gates of Regensburg. Today, the grave of the world famous astronomer cannot be found. Only a few years after his death, the graveyard was destroyed during the siege of Regensburg by Bernard of Weimar and the subsequent recapture by Bavarian troops. The inscription on his tombstone read:

> My spirit strode across the heavens, now I measure the depth of the Earth. Heaven gave me my spirit, here now my earthly body rests.

Chapter 10
The Order of Things Revisited

We have examined the motivations which induced Johannes Kepler to travel the way he did, both physically and spiritually, finally consummated in the legacy he left behind in his major works. Kepler himself was convinced that he had deciphered God's blueprint for the harmony of his creation. In this respect, Kepler stood at the summit of all the combined efforts of those that preceded him in the history of science. And to this day, he has not been surpassed by any other. For all practical purposes, however, his tangible legacy remains the compilation of the Tabulae Rudolphinae and his three laws of planetary motion. Everything else would soon have to be abandoned.

Mach and Others

Johannes Kepler designed his ultimate harmony within the closed space of our planetary system, neglecting the rest of the world beyond—except for a descriptive component for the fixed stars. The latter took on the role of "onlookers", to use today's terms. The other condition required for Kepler's system to function was the presupposition of absolute space and absolute time, although Kepler himself did not postulate it as such, but clearly took it for granted. Newton was convinced that he could prove this by a thought experiment using a bucket. He concluded that one could always see when water in a bucket was rotating around its rotation axis relative to absolute space, since in this case the surface of the water would form a rotational paraboloid caused by the centrifugal forces—independently of whether the bucket itself were also rotating.

This argument did not hold for ever. It was first refuted by George Berkeley, an Anglican theologian and philosopher. His argumentation was later adopted by Ernst Mach in 1883, an Austrian physicist and philosopher (Fig. 10.1). Mach argued that Newton neglected the influence of all other matter in the Universe. Newton's conclusion would only hold in an otherwise empty Universe. This later became known as Mach's principle, and it was formulated in many different ways—among others by Albert Einstein in his general theory of relativity, although it turned out in

© Springer Nature Switzerland AG 2020
W. Osterhage, *Johannes Kepler*, Springer Biographies,
https://doi.org/10.1007/978-3-030-46858-3_10

Fig. 10.1 Ernst Mach

the end that these were not compatible either. What is left is the assumption that all celestial bodies, wherever they reside, exert an influence on all others, and so have to be taken into account.

Newton

The motions of celestial bodies could be described with some accuracy, but the explanation for these motions had to wait for the genius of Isaac Newton.

Newton was born in 1642, the year when the civil war between the crown and parliament broke out in his country. In the year of his death, 1726, Jonathan Swif t published his novel "Gulliver's Travels". Newton's most relevant work for this chapter was "De Motu Corporum" ("On the Motion of Bodies"), published in 1684. In this publication, Newton summarized his own mechanical experiments. His practical and theoretical results were later incorporated in "Philosophiae Naturalis Principia Mathematica" ("Mathematical Principles of Natural Philosophy"). In this major work he consolidated the results of Galileo's experiments on motion, Kepler's observation of the motions of planets, and reflexions on inertia by Descartes. His three laws of gravitation formed the foundation of classical mechanics:

1. A body remains in a state of rest or uniform straight-line motion if it is not acted upon by any forces to change its state.
2. Any change in motion is proportional to the force acting and proceeds along the straight line in which the force acts.
3. Equality of action and reaction; more precisely, the actions of two bodies upon each other are always equal and in opposite directions.

In a sense, with Kepler, we reached the summit of order, in the form of a perfect cosmic harmony. Newton now dissolved it. By his consistent application of the laws of gravity, he accepted the mutual influence of all celestial bodies upon one another, finally resulting in many-body problems whose exact calculation is only possible using numerical methods and powerful computers, even today. Because of this mutual influence, the orbits of the planets become idealised trajectories which deviate in reality due to interference from other planets. Reality now appears too complex to allow for perfect cosmic order.

The calculation of perturbations is an important application of mathematics to celestial mechanics. It was triggered in 1820 by the discovery that the orbit of Uranus deviated from its calculated path. The French astronomer Urbain Le Verrier applied it in 1844 to calculate the orbit of an unknown planet that might explain the deviations in Uranus' orbit. Two years later, the German astronomer Johann Gottfried Galle discovered the planet Neptune on this basis. At the beginning of the 20th century, the American astronomers Percival Lowell and William Henry Pickering applied perturbation calculations to calculate the orbit of Pluto, which was only discovered decades later by the Lowell Observatory.

Thus a perturbation of the path of a celestial body amounts to a deviation from a calculated course. One can distinguish between irregular perturbations and periodic perturbations, which describe fluctuations around some average value, and secular perturbations, representing long term monotonic changes. Various things may be responsible for these perturbations, including for example gravitational forces caused by changes in a gravitational field or the influence of other celestial bodies, but also relativistic effects like time dilation or the curvature of space. To describe the orbits of asteroids, the gravitational forces exerted by the Sun, all the planets, the Moon, and other large celestial bodies have to be taken into account. For some Earth-orbiting satellites, perturbations due to the irregular gravitational field of the Earth play an important role. Periodic perturbations are also caused by the influence of the Moon and Sun, similar to the tides.

Even today, it is still not completely understood how our planetary system came into being some 4.5 billion years ago as a system with the stable planetary orbits as we observe them today. This means that Newton's theory is not yet able to explain what Kepler's nested polyhedron model could: the mean distances between the Sun and Mercury, Venus, Earth, Mars, Jupiter, and Saturn.

But today, creation and harmony are two different things altogether. The notion of a "Big Bang" itself excludes any association with harmony as such. Newton himself laid the foundation for a different kind of cosmology than Kepler's with his laws of gravitation. He was the first to express the conviction that the Universe would be homogeneous and isotropic, i.e., that the Universe would present the same view to an observer wherever he might be situated—with the exception of some local irregularities. The German philosopher Immanuel Kant took Newton's conception of space, time, and gravitation to develop a cosmological model starting out from some primordial chaos. Gravitational collapse was avoided by some kind of repulsive force between the planets.

The End of Certainty

By the end of the 19th century the systematic exploration of the directly discernible world had been accomplished by successfully verifiable theories. Space in the immediate vicinity of Earth was understood on the basis of Newtonian-Keplerian mechanics. In 1899, the head of the US Patent Office gave up his job, because according to his insights there would be no further significant innovations. The physicist Albert Michelson was of the opinion that the only thing left to do in science at the end of the 19th century was just to refine what was already known, since there would be no novelties to be discovered. Philipp von Joly, professor of physics at Munich University, when asked by the young Max Planck, advised him that nearly everything had already been explored in physics, and the task at hand was to close a few unimportant gaps.

The following steps were triggered by chance discoveries.The road into the microcosm was soon to be opened. Conrad Roentgen had discovered a new type of radiation, and this encouraged Henri Becquerel to investigate further fluorescent substances. It was by pure chance that he discovered radioactivity, because he had stored uranium salts on top of sealed photographic plates. Three types of radiation are known from nuclear decay: alpha rays (${}_2H^4$), beta rays (e^-), and gamma rays (electromagnetic waves). Two years after Becquerel, the Curie s discovered two other radioactive substances: radium and polonium. Joseph J. Thomson discovered the first elementary particle, the electron, and developed a first theory of the atom. Ernest Rutherford classified radioactive radiation and developed a new atomic model with a nucleus.

Then, in 1900, Max Planck gave his lecture to the German Physical Society (Deutsche Physikalische Gesellschaft) that would reshuffle the whole pack once again.

Max Planck

Max Karl Ernst Ludwig Planck was born on April 23rd 1858 in Kiel and died on October 4th 1947 in Gottingen. In 1878, he accomplished his state examination for a teaching post in higher education for the subjects mathematics and physics in Munich. In 1880, he became associate professor at Munich University, in 1885, associate professor of theoretical physics at the Christian Albrechts University in Kiel, and in 1892, professor of theoretical physics in Berlin. From 1912, he was appointed permanent secretary of the newly founded Kaiser Wilhelm Society for the Advancement of Science (Kaiser-Wilhelm-Gesellschaft zur Förderung der Wissenschaften) (later, the Max Planck Gesellschaft). In 1914, he became president of the university in Berlin.

Planck's most important finding was the discovery that the spectral energy distribution of a black body is not continuous, but discrete. The emitted waves correspond to quantized energy states. One of the immediate consequences was that this demanded a rethink. For one thing, the continuum had been lost, i.e., events in nature must sometimes be described by discrete states. For another, certainty had been lost,,replaced now by probability. This was the birth of quantum physics.

Albert Einstein

Quantum theory thus brought into question concepts which until then had been dear to Western ways of reasoning, not only in science but even in everyday life. However, there was more to come, or rather to lose. The cause for this was another theory that would spring up only five years after Planck's seminal lecture: the theory of relativity.

Albert Einstein was born on March 14th 1879 in Ulm and died on April 18th 1955 in Princeton. In 1900, he obtained a teaching diploma in mathematics and physics at the polytechnic in Zurich. His first employment was as a tutor in Winterthur, Schaffhausen, and Bern. In 1902, he was engaged by the Swiss Patent Office in Bern as Technical Expert 3rd Category. Then came the year of grace. In 1905, he published four major papers:

- "Ueber einen die Erzeugung und Verwandlung des Lichts betreffenden heuristischen Gesichtspunkt zum photoelektrischen Effekt"(About a Heuristic Aspect Regarding the Photo-Electric Effect Concerning the Creation and Transformation of Light)—For this Einstein was rewarded the Nobel Prize in Physics later.
- "Eine neue Bestimmung der Molekueldimensionen" (A New Determination of Molecular Dimensions)
- "Ueber die von der molekularkinetischen Theorie der Wärme geforderte Bewegung von in ruhenden Fluessigkeiten suspendierten Teilchen zur Brownschen Molekularbewegung" (On the Motion of Small Particles Suspended in Liquids at Rest Predicted by the Molecular Kinetic Theory of Heat)
- "Zur Elektrodynamik bewegter Koerper" (On the Electrodynamics of Moving Bodies)—This was the foundation of the special theory of relativity.

In 1913, Einstein became full-time paid member of the Prussian Academy of Science (Preussische Akademie der Wissenschaften) in Berlin, and in 1914, Director of the Kaiser Wilhelm Institute for Physics. In 1916, he published his general theory of relativity. From 1933 onwards, he worked in the Institute for Advanced Study in Princeton. Until the end of his life, he tried to develop a unified field theory of the forces in nature—without success.

In his 1905 paper about special relativity, he explained the results of the Michelson-Morley experiment. His conclusion was that the speed of light in vacuum is independent of the motions of the light source and the observer. It is a natural constant with the value c. Furthermore, he insisted that all laws of physics have equal validity in every inertial system. This meant that inertial systems were basically not distinguishable from one another. The most important finding, however, was that there exist neither absolute space nor absolute time. This is corroborated by the following three phenomena:

- The relativity of simultaneity
- Time dilation
- Length contraction

and .... $E = mc^2$.

The principle of general relativity reads as follows:

If the same laws of physics are valid in two systems, then there exists no reference system for absolute acceleration, just as in special relativity there is no absolute velocity.

Conversely:

When physical laws are valid in one environment, they are equally valid in any environment moving relative to the first.

This principle of equivalence can already be found in Newton's mechanics, where inertial mass is equal to passive gravitational mass. Einstein assumed that the principle of equivalence is not only valid in mechanics. He stated that there exists no experiment whatsoever capable of distinguishing between uniformly accelerating motion and the presence of a gravitational field.

The main consequence is that the curvature of space is proportional to the distribution of energy and momentum. The theoretical results of the general theory of relativity (GTR) have led to a revolution in philosophy and natural science. In the end, the cosmos, the Universe, has once again given cause for the construction of entirely new models.

During the first half of the 20th century, the limits of both microcosm and macrocosm were significantly extended. During the second half of the 20th century, research in physics arrived at the limits of human perception: on one side, the Universe and its creation, and on the other, the smallest particles, the quark s. Theory became more and more complex and at the same time fundamentally enlarged the modern world view.

Subjectivity and Objectivity

If we approach the notion of "truth" from a philosophical perspective, matters become rather confusing concerning the possibilities for getting close to it. One can start with Augustinus or even with Plato, or ask the scholastics or later Kant—indeed, one could navigate through the whole of European intellectual history. At some stage there comes the jump into post-modernity, which has found the truth that there is no objective truth.

At the beginning of modern times, natural science could still counter such a claim by arguing that it disposes of an objective approach to reality—equating reality and truth for the purpose of simplification. But even if we assume this, since the advent of quantum theory, science has become aware that it no longer has access to a complete reality. The intervention of an observer changes reality through his observation, and thus also changes the perceived truth of the matter at the same time. This is no speculation, but the result of experiments.

Basically, this insight leads us back to the early times of human reflection, when for the first time the separation of a subject from his environment took place in his conscience. In the end, truth can only be found in external objectivity, absorbed internally by a subject, and then interpreted. This means that the whole truth can never be captured by reception of external truth free of doubt, but always only as an approximation via subjective means. The whole truth thus comprises subject and object, in a manner of speaking.

Taking this dilemma between subjectivity and objectivity, the theologian Dieter Hattrup formulated a thesis which may be of some use in our further discussion [29]. It reads:

Truth and self-interest stand opposite to each other.

This is based on the following hypotheses:

- Truth and self-interest are in constant competition.
- Truth will be found nowhere unless is serves some useful purpose.

Against this background we can now leave the philosophical realm and exchange it for the everyday world, because

- if truth is reduced to self-interest, talk about truth will always create an aura of suspicion, and
- in everyday life, truth and self-interest cannot possibly be separated from each other, although they can be distinguished in thought.

Thus truth has to do with communication (talk) in the widest sense, but not exclusively. On the other hand, communication is always connected to truth or the lack of it. Truth itself is tolerant, but to recognize it, a space of freedom is required, in which it is possible to correct fallacy.

Table 10.1 The four forces of nature

Force	Field quantum	Relative strength	Range [m]
Strong interaction	Gluon	1 (reference)	10^{-15}
Weak interaction	Vector boson	10^{-1}	10^{-18}
Electromagnetism	Photon	10^{-2}	∞
Gravitation	Graviton	10^{-39}	∞

Deconstruction

In classical terms, not much was left of Kepler's legacy (except his three laws as classical approximations). The harmony of the cosmos was torn up in the first half of the 20th century. But the sources of motivation that lies in the idea of constructing or finding the order of things was still there. The task at hand was now to reassemble the scattered pieces of the world and bind them together into a new and harmonious construct, thereby fulfilling the prophecy of John A. Wheeler: "Someday a door will surely open and expose the glittering central mechanism of the world in its beauty and simplicity".

Four fundamental forces have been identified in nature: gravitation, electromagnetism, the weak interaction, and the strong interaction. Electricity and magnetism were unified in the 19th century by James Clerk Maxwell. Weak interaction and electromagnetism were formally unified as the electroweak interaction by Abdus Salam and Steven Weinberg in the 1960s. Together with the quark model and the associated theory of quantum chromodynamics, all this was put together to form the Standard Model of particle physics. What was left aside in this model was gravitation. It plays a major role in another standard model, the cosmological one, based on the general theory of relativity. Table 10.1 compares the four forces of nature.

In modern physics, the term "interaction" is used instead of "force". Force only has meaning in classical physics. Field theory teaches us that interactions are exchanged via field quanta.

To sum up:

(1) Electromagnetic interaction:
The field quantum is the photon.

(2) Strong interaction:
The field quantum is the gluon, holding together the quarks in the atomic nucleus.

(3) Weak interaction:
At low energies, this is much weaker than the electromagnetic interaction, and even more so with respect to the strong interaction. It is responsible for beta-decay. Its field quanta are the W- and Z-bosons.

(4) Gravitation:
This is much weaker than the three other interactions. The gravitational attraction between a proton and an electron is 10^{-39} of the electrical attraction.

The Standard Model explains all phenomena observed in nature—with the exception of those caused by gravitation. This opens up a parallel physical world, leading its own life alongside or even counter to quantum physics. There have been many attempts to bridge the gap—without success up to now.

Throughout this book, gravitation has been the overriding cause for what Kepler and his contemporaries were trying to describe, although the term and its meaning were not known at the time. Today, every aspect of gravitation is based on Einstein's general theory of relativity. But what we are talking about here is more than just a theory about a force of nature. Gravitation or the general theory of relativity opens the door to cosmology and the origins of the Universe.

At present, we find ourselves in the following situation:

1. Atomic physics is an established science, the fundamentals of which are well understood. Today, its subjects of interest are the precise measurements of energy levels to determine certain natural constants even better using laser technology, individual investigations of ions, and spectroscopic analyses of specific materials.
2. Nuclear physics has had—as we all know—a varied history outside basic research. Evidence for the correctness of the theoretical basis has been obtained in the military sector as well as in the civil sector. A continuing field of basic research, apart from ongoing technological improvements, is electron scattering to find out more about nuclear structure.
3. High energy physics and elementary particle physics are a long way from reaching their culmination. This is borne out by continuing experiments with more and more powerful accelerators.
4. Modern cosmological observations by a range of space-based optical and radio telescopes are continually broadening our view of the cosmos, leading to a constant verification of cosmological models and the validity of the theory of relativity.
5. There are two things, which have still not been achieved to this day:

- The unification of the four known forces in nature
- The synthesis of quantum theory and the theory of relativity.

A new cosmic harmony thus still seems a long way off, 400 years after Kepler.

Riddles

We should distinguish between unresolved problems and open questions, which are currently being attended to, and mysteries, whose solutions remain in the realm of speculation at present. There exist transitions between these categories. Here are some examples:

- String theory
- Cosmological constant and dark matter
- TOE: the theory of everything, unifying all interactions
- The Big Crunch.

Copernicanism implied the removal of man from the centre of our planetary system, then the removal of the Sun from the centre of the world, then the removal of our galaxy from the centre of the Universe. The final step would be the assumption that even our universe is not unique. The multiverse would be the final insult to Johannes Kepler.

There exist a number of open questions which have led to the concept of the multiverse—a world comprising an infinite number of parallel worlds. The trigger for it came from quantum physics and its superpositions of quantum state s, which are only resolved when a measurement takes place, for example. According to the type of measurement, a distinct branch of many possible states is opened up: one tangible universe among a possible multitude. This idea of quantum physics interpretation was first put forward by Hugh Everett.

The creation of a multiverse could be explained by the hypothesis of the inflationary expansion of the Universe shortly after the Big Bang (after 10^{-35} to 10^{-33} s) by Alan H. Guth. This hypothesis answers several cosmological questions at the same time, e.g., regarding the homogeneity of the Universe and the cosmic background radiation, but also the missing curvature of the cosmos and the density fluctuations arising from quantum fluctuations in the initial inflationary field and leading eventually to galaxy formation.

One consequence of these assumptions is that expanding bubbles would be created which could not communicate with each other. Andrei D. Linde postulated just such a bubble theory and thus one form of multiverse. The existence of a multiverse would explain the fine adjustment needed to enable the existence of intelligent life in the cosmos, even if one assumes that in other parallel universes such a fine adjustment does not exist. Because all potential universes are possible, a universe like ours must inevitably exist. This would solve the problem of the Anthropic Principle.

The theory of the multiverse is still only a theory. A proof has not been delivered yet and indeed cannot be delivered according to the framework of the theory itself, since there is no way to receive or send signals from one parallel world to another. Otherwise they would not be parallel worlds, but only parts of one and the same universe.

Rebuilding Harmony

The search for a Grand Unified Theory goes on. The unification would comprise the electromagnetic, strong, and weak interactions. It assumes that these three forces were equal shortly after the Big Bang, i.e., unified in a single force. One way to test this assumption would be to detect and study the decay of the proton, but such a decay has never been observed. All we have is an upper limit on its half-life, which corresponds to approximately the age of the Universe.

Albert Einstein was already looking for a unified field theory, albeit only for the unification of gravitation and electromagnetism, since the other two forces of nature were not known then. He tried for more than 30 years to achieve this but in the end failed. As early as 1926, he had declared in his acceptance speech for the Nobel Prize: "The mind striving after unification of the theory cannot be satisfied that two fields should exist which, by their nature, are quite independent."

Although he was already contemplating this unified field theory at that time, he recommenced this work more seriously in Princeton in 1945. Until 1951, he was assisted by his old friend from their Zurich years, Hermann Weyl. Einstein did not want to solve the quantum problem by first developing a unified classical theory and then merge it with quantum theory. He wanted his theory to be capable to reproduce the results of quantum theory in the first place. But his principle of equivalence only showed him how gravitation could be identified with the structure of space-time. He failed in his search for a physical principle which would be the basis for a geometrical description of all interacting forces, and thus for the creation of a unified geometrical field theory. One of the reasons could have been his lack of understanding of the quantum world, as the physicist Brian Green thought: "Einstein did not know enough about the fundamental processes and principles of the world of microscopically small particles to achieve the next step."

Another approach would be to unify gravitation and electromagnetism in the context of relativity and later return to the quantum question [30]. The first step could be achieved by establishing two fundamental things:

- Constructing a five-dimensional manifold with three space coordinates and one time coordinate, the fifth coordinate being represented by an affine parameter q/m (charge/mass)
- Formulating a higher principle of equivalence:

> If an observer in a (special) inertial frame (i.e., a Faraday cage) is travelling in an 'accelerating' field, he can neither distinguish whether he is moving in a gravitational or an electromagnetic or any other field, nor whether he is subjected to an apparent field (of any kind) due to acceleration of a reference frame.

Then criteria for a true unification between gravitation and electromagnetism would be the following:

a. The two interactions have to be described by the same field equation.
b. On a geometric basis, all forces have to be replaced by a coupling of a single main source term to geometry.
c. All dimensions of the 5-dimensional manifold must be in some way physically observable.

The gravitational force is replaced by the coupling of the curvature of a space-time manifold to an energy source term. As a consequence, the electromagnetic "force" must also be replaced by the coupling of a manifold curvature to an energy source term. Both the manifold in which the action takes place and the source have to be common to gravitational and electromagnetic events.

Whatever the final outcome of all the current attempts to formulate GUT, there would be no reason to stop there. The next step would lead to TOE, a Theory of Everything, including all four forces or interactions in nature and thus bridging the gap between quantum theory and the theory of relativity. This world formula has not been found yet. It is doubtful whether it could ever be found by adhering to these two branches of modern physics, quantum theory and the theory of relativity. Indeed, a totally different approach may well be necessary.

The term "world formula" was first coined by Werner Heisenberg ("Weltformel"). Much earlier people were talking about "Machina mundi" or world machine. There is one person, who—until very recently—devoted much effort to finding such a mechanism. In the middle of the 1970s, Stephen Hawking proclaimed that he would succeed in finding a TOE in about ten years time—in any case not later than twenty years after his announcement. Then, after another twenty years, he said, that the projected twenty years had just begun. Hawking was looking for this in the framework of string theory. This theory claims that an elementary particle is something like a string. The different states of excitation of a string represent different elementary particles. Strings are 10^{25} times smaller than an atom, and they oscillate in a ten-dimensional space-time. Some string theoretical models include quantum gravity, so gravitation may finally be included in a common world harmony.

Hawking's efforts to find a TOE were by no means the first attempt.

The deterministic "Laplace Demon" was an approach to construct a TOE on the basis of Newton's mechanics. Pierre-Simon Laplace mentioned this construct in the preface of his "Essai philosophique sur les probabilities" (Philosophical Essay on Probabilities) in 1814. His demon would be intelligent, calculating all acting forces in the cosmos as well as all positions and velocities of all bodies. According to the laws of Newtonian mechanics, this being would be in a position to calculate the development of the Universe both forwards and backwards.

Heisenberg also believed he was on the point of making the final breakthrough and being able to formulate a TOE. He gave a radio interview in 1958 in which he claimed that he and Wolfgang Pauli (Fig. 10.2) would shortly publish a world formula based on the spinor theory of elementary particles. Only some minor technical details had still to be attended to. But Pauli left the project because he found that it was not feasible. Pauli wrote a letter to the Russian-born physicist George Gamow. It contained an empty rectangle and the following comment: "This is to demonstrate

Fig. 10.2 Wolfgang Pauli

to the world that I can paint just like Titian. There are only some technical details missing" [25].

Pauli himself wrote an essay about Kepler in which he expanded on his understanding of a world formula. For Pauli, as for Kepler, a world harmony or a Machina mundi was an ordering principle in nature, society, music, and other areas. He used the term "archetype" as a metaphor for primary mathematical intuition.

The generally accepted cosmological model today is the "standard hot Big Bang model", based on the assumption that, although gravitation dominates the overall development of the Universe, the observed details are determined by the laws of thermodynamics, hydrodynamics, atomic physics, nuclear physics, and high energy physics.

It is assumed that during the first second after the beginning, the temperature was so high that a perfect equilibrium prevailed between photons, neutrinos, electrons, positrons, neutrons, protons, diverse hyperons and mesons, and possibly gravitons. After several seconds, the temperature fell to around 10^{10} K and the density amounted to around 10^5 [gcm^{-3}]. Particles and antiparticles annihilated in pairs, hyperons and mesons decayed, and neutrinos and gravitons decoupled from matter. The Universe now consisted of free neutrinos and perhaps gravitons. In the following period between 2 and 1000 s after the beginning, a first primordial creation of elements took place. Before that such attempts had been destroyed by high energy protons. These elements were basically alpha particles (He^4), traces of deuterium, He^3, and Li. These made up for 25% and the remainder was hydrogen nuclei (protons). All heavier elements were created later.

Between 1000 s and 10^5 years after the beginning, thermal equilibrium was conserved by a continuous transfer of radiation into matter as well as by permanent ionization processes and the building of atoms. Toward the end the temperature fell to a few thousand degrees. Thereafter the Universe became dominated by matter rather than radiation. Photons had no longer enough energy to ionize hydrogen atoms permanently, for example. After the photon pressure had disappeared, matter could begin to condense into stars and galaxies, so from 10^8 to 10^9 years after the beginning. There is still uncertainty about the cause of those little disturbances which in the end violated the perfect initial isotropy to enable the formation of these different structures.

The calculations based on gravity in this model use simplified assumptions. First of all galaxies are treated as gas particles. At the same time, the internal structure of these particles is neglected. To simplify the calculations even more, the gas is assumed to be an ideal liquid. This gas is characterized by a velocity vector field, a mass energy density, and a pressure. This leads to an energy density tensor for this cosmic liquid.

Of course, there are modified versions of this model, but we shall not discuss those any further here. They include quantum fluctuations, inflationary expansion, percolation theory for the creation of hadrons from a quark-gluon plasma, etc.

Concerning the above, Harald Boettger cautions with regard to the efficiency of physics in general and TOE in particular [25]. He quotes the physicist Paul Dirac, who states that the researching physicist, having made a discovery, should be concerned about defending his present point of view and looking out over the field before him. His question is: Where do we go from now? What are the consequences of these new discoveries for the rest of science? How far can we illuminate other problems lying before us with these new insights? He also quotes Einstein, who had been asked whether everything could be reproduced scientifically. Einstein's answer was that this seemed possible, but that it would not make any sense at all. It would result in a reproduction by inadequate means, like representing a Beethoven symphony by an air pressure graph."

Boettger concludes that there are narrow limits for the scope and usefulness of TOE—even in physics itself. Even Stephen Hawking no longer believed in the possibility of TOE in the end, because of Gödel's incompleteness theorem for formal systems, according to which no part of a system can produce a conclusion about the whole. So man, as part of nature, will not be able to draw conclusions about nature as a whole.

The search for harmony continues.

What About the Music?

Kepler's "Harmonices Mundi" is largely a book about music (the geometrical part is a repetition from "Mysterium Cosmographicum"). The music of the spheres had been replaced by the non-audible acoustic harmony generated by the orbiting planets. The chords were derived from the ratio between the maximum and minimum speeds on their elliptical paths. No one nowadays pays any attention to such assumptions. Or so it would appear.

Is there any nexus between physics and music? Of course, there is. Acoustics is a physics discipline in its own right. The generation of sound and music can be explained by physical equations. But can mathematics be applied to musical works of art like symphonies or even simple songs? Would people still care about a transcription of music into mathematical language and use such mappings as a tool for classification? The answer is: yes. Even 400 years after Kepler, the subject has not been closed.

In 2002, C. Hartfeldt et al. published an essay about the role of mathematics in music [31]. In fewer than 60 pages, they uncovered the mathematical secrets of compositions. They started out with an appreciation of Kepler's music of the planets and a critique of Arnold Schoenberg's harmonics, before diving into more detail. One such detail was the golden section in music. In their review, the authors referred to Fibonacci series, interval proportions, and frequency relations, and even discussed the relationship between geometrical and musical structures.

Another approach to classifying music can be found in statistics. Wilhelm Fucks published a paper in 1962 based on the analysis of music including random sequences, thus relating music and chance [32]. In his analysis, he treated the sequence of musical elements and intervals in a composition as just a random series of values which can then be analyzed statistically in terms of frequency distribution or cross- or auto-correlations. In the end, it is possible to find characteristic values for specific composers. But far from reconstructing a new musical harmony, this approach was another attempt at deconstruction.

The Limits of the World

Let us conclude this chapter with a quote from the French philosopher Michel Foucault:

> And it is here that we find that only too well-known category, the microcosm coming into play. This ancient notion was no doubt revived, during the Middle Ages and at the beginning of the Renaissance, by a certain neo-Platonist tradition. But by the sixteenth century it had come to play a fundamental role in the field of knowledge. It hardly matters whether it was or was not, as was once claimed, a world view or "Weltanschauung". The fact is that it had one, or rather two, precise functions in the epistemological configuration of this period. As a "category of thought", it applies the interplay of duplicated resemblances to all the realms of nature; it provides all investigation with an assurance that everything will find its mirror and its macrocosmic justification on another and larger scale; it affirms, inversely, that the visible order of the highest spheres will be found reflected in the darkest depths of the Earth. But, understood as a "general configuration" of nature, it poses real and, as it were, tangible limits to the indefatigable to-and fro of similitudes relieving one another. It indicates that there exists a grater world, and that its perimeter defines the limit of all created things; that at the far extremity of this great world there exists a privileged creation which reproduces, within its restricted dimensions, the immense order of the heavens, the stars, the mountains, rivers, and storms; and that it is between the effective limits of this constituent analogy that the interplay of resemblances takes place. By this very fact, however immense the distance from microcosm to macrocosm may be, it cannot be infinite; the beings that reside within it may be extremely numerous, but in the end they can be counted; and, consequently, the

similitudes that, through the action of the signs they require, always rest one upon another, can cease their endless flight. They have a perfectly close domain to support and buttress them. Nature, like the interplay of signs and resemblances, is closed in upon itself in conformity with the duplicated form of the cosmos. [2]

Chapter 11
Conclusions

Johannes Kepler, although not poor in a general sense, especially if we apply this criterion to the living conditions of the ordinary people of his time, never became a wealthy person either. He had been affected by various illnesses from early childhood and was forced to lead an uncertain life in which no place on Earth could be called home. When he was a child, this was the result of the continuous movements of his parents, but later it was brought about by the turmoil of religious strife culminating in the Thirty Years War. At the same time, he suffered the early death of his first wife and some of his young children, although this again was not uncommon in a time when infant mortality was generally high. Against this backdrop (or because of it?), his achievements are to be judged as all the more extraordinary.

He became famous all over Europe during his lifetime as an astronomer, and his advice was sought in many places. He corresponded with important peers such as Galileo and Bernegger. In fact, his fame and accomplishments led to special treatment by various authorities even during the hard times of the Counter-Reformation. The Tabulae Rudolphinae and his three astronomical laws served as the basis for Newton to develop his own theory, laid down in his "Philosophiae naturalis principia mathematica". Kepler's thinking played a central role in the natural philosophy of Friedrich Wilhelm Schelling and Georg Wilhelm Friedrich Hegel. He continued to be regarded as the genius *par excellence* well into the time of romanticism. But by the middle of the nineteenth century, Kepler's works and methods became more and more subject to criticism from empiricism and positivism.

Among Kepler's outstanding qualities was his ability for lateral thinking. The term "lateral thinking" today is sometimes misused to depict a person as someone who is at odds with traditional thought and develops a new and radical approach to existing challenges. However, the original meaning of lateral thinking refers primarily to the ability to approach a given problem in multiple fashions, i.e., by using methods of analysis and synthesis of a number of different scientific disciplines, and by synthesizing these fragmented results into a coherent overall model. This was Kepler's strongest point: looking through the stars and seeing geometrical shapes, and listening to cosmic music, while at the same time being submerged in theology and mysticism. Neither before nor after has anyone endeavored a similar approach

© Springer Nature Switzerland AG 2020

W. Osterhage, *Johannes Kepler*, Springer Biographies,
https://doi.org/10.1007/978-3-030-46858-3_11

nor succeeded in building anything like this when trying to describe the world. In terms of his ability as a lateral thinker, in recent times maybe only John Nash came close to Kepler's capabilities.

John Nash

The mathematician John Nash (Fig. 11.1) did not contribute directly to astronomy, but he is one of the rare examples of a true lateral thinker. He possessed the ability to connect a variety of mental threads to form concise solutions of problems encompassing economic science, game theory, and strategy, and at the same time created byproducts with consequences for the description of space-time. His latter work on manifolds led to concepts in algebraic geometry proving that Riemannian manifolds as employed in general relativity could be embedded in Euclidean space. Despite the fact that he suffered from paranoid schizophrenia (or because of it?), he made a vast contribution to differential geometry.

Fig. 11.1 John Nash

Kepler's Legacy

There is no doubt about the validity and importance of Kepler's three laws in astronomy, which can be found in every textbook about astronomy today. But the Tabulae Rudolphinae, his final great achievement, also constituted a major contribution to practical astronomy. Until the 18th century, they served as the basis for most astronomical calculations. Isaac Newton used them to formulate his theory of the gravitational force. And Adam Schall von Bell, a German Jesuit, who worked at the court of the Emperor of China Xu Guangqi until his death in 1666, completed the reform of the Chinese calendar, which had been started earlier by Johannes Schreck, another German Jesuit and missionary, with the help of these tables.

Kepler left a vast number of publications, a list of which can be found in the next chapter. Among the more scientific ones, rather than those bordering on mysticism, is the one called Dioptrice, in which he outlined the foundations of optics. He explained the refraction of light and optical imaging without knowledge of the laws of diffraction later developed by Willibrord van Roijen Snell, a Dutch astronomer and mathematician, first published, however, by his countryman Christiaan Huygens in 1703. As a byproduct, he designed an improvement of the Galilean telescope, built in 1611 using two collecting lenses and producing an inverted laterally reversed image. Today, a space-based telescope called "Kepler", built by NASA and launched in 2009, is in operation to detect exoplanets orbiting around distant stars, mainly in the region of the constellation of Cygnus (Fig. 11.2). Naturally, it uses far superior techniques than were known during Johannes Kepler's lifetime, but it still applies his laws. It has found more than 2300 exoplanets up to now.

Regarding the remainder of Kepler's work, it was once again Wolfgang Pauli who wrote in his essay "Der Einfluss archetypischer Vorstellungen auf die Bildung naturwissenschaftlicher Theorien bei Kepler" (The Effect of Archetypical Conceptions on Kepler's Development of Scientific Theories): "Theories are not derived by forcing conclusions out of formal records. Theories come about by understanding inspired by empirical material. The process of understanding nature seems to rest on the matching of pre-existing internal images with external objects and their behavior" [25]—just as it may have happened for Johannes Kepler. But this implies in essence that he derived his results by other means than causal reasoning. This is not surprising, since causality only asserted itself in the exact sciences from the 18th century onwards, starting with Newton's mechanics.

If Kepler had been trying to find some sort of theory of everything, we may wonder how far such concepts could go. Obviously, his understanding of harmony went beyond physics as such, since it comprised mathematics (geometry) and music, and all this influenced by theology. Theology in its own right has always claimed to encompass everything connected with human existence, thus standing for something like a super theory of everything. But quantum physics and physics in its reductionist forms make the same claim—even though they have not yet achieved this goal—contrary to theology.

Fig. 11.2 Kepler is mounted on top of the third stage of its Delta II 7925 launcher. *Source* NASA

Another remaining source of ambiguity is Kepler's relation to astrology. Although he made a living from his prognostica and horoscopes, he changed his point of view several times in his lifetime concerning the scientific validity of this discipline. In his publication Tertius Interveniens, he defended astrology against attacks from the physician Philip Fresenius. He conceded that many professional astrologists were cheats, but claimed that that should be no reason to condemn astrology as a whole. In his De fundamentis—a prognosticum for the year 1602—he tried to put astrology on a solid scientific basis in 51 propositions. After criticizing Aristotle and claiming the existence of a totally animated nature, he distinguished different types of "souls" that were receptive to different influences. And then he proclaimed the existence of a kind of soul that was capable of perceiving the geometrical and harmonious structure of the cosmos. One important point he made was the symmetry between God and his created world, thus elevating God to the rank of geometrician himself. In the end,

Johannes Kepler was convinced about the scientific value of astrology, but rejected the way it was traditionally practised, with its elements of superstition.

Back to Square One?

On the 30th of January 2019 an article appeared in the Frankfurter Allgemeine Zeitung entitled "A New View of the Galactic Monster—The Beam of Matter from the Galactic Black Hole Seems to be Directed toward the Earth". It reads as follows (translated by the author):

With the help of an array of radiotelescopes, astronomers have taken a detailed look at the region surrounding the black hole at the centre of the Milky Way. The findings of the scientist Sara Issaoun at Radboud University in Nijmwegen in the Netherlands confirm what observers with optical telescopes have recently suggested: we appear to be looking directly down a beam of matter in which gaseous masses are accelerated up to near the speed of light. The work of Issaoun and her team, comprising researchers from the Max Planck Institute in Bonn, have been published in the Astrophysical Journal.

Astronomers assume that every big galaxy hides a black hole in its centre. In the case of the Milky Way, the massive monster is estimated to weigh as much as four million times the Sun. From a distance of about 26,000 light years, it appears in our sky not bigger than a tennis ball on the Moon. Hence, astronomers combined a dozen radiotelescopes distributed all over the globe, among them the radio telescope in Effelsberg in Bad Munstereifel in order to carry out Very Long Baseline Interferometry (VLBI). Because of the great distance between them, the telescopes can produce a view as sharp as with a single telescope the size of the Earth. For the first time, researchers included the Atacama Large Millimeter Array (ALMA) in the Atacama desert in Chile. ALMA is the most sensitive radiotelescope in the world in the frequency range of 86 Gigahertz. Since it is the only telescope in the configuration placed in the southern hemisphere, its use doubled the angular resolution of the configuration.

The reconstructed images do not show the black hole itself, but a diffuse fog of radio beams. This is created around the hole: in its vicinity gas and dust gather into a rotating disc, before they are drawn into the hole from time to time. A fraction of the matter escapes in two opposite radiation cones, called jets, oriented perpendicular to the disc. According to astronomers, these jets are the source of the radio frequency radiation emanating from the black hole. According to the VLBI measurements, the radiation has a compact and symmetrical structure best explained by the fact that one of the jets is directly oriented towards the Earth.

Astronomers would very much like to penetrate this fog and thus come even closer to the black hole. With the help of their Event Horizon Telescope (EHT), they are currently trying to delineate the blackness of the hole in front of the background of the diffuse radio radiation and thus visualize its event horizon, i.e., the limit beyond which no escape is possible because of its gravitation. The EHT also uses VLBI techniques, but with higher frequencies. The first results are expected in spring 2020.

On the 9th of February 2019, a comment on this matter appeared as a letter to the editor of the Generalanzeiger Bonn by Prof. emeritus Wolfgang Kundt, a renowned astrophysicist at the Friedrich Wilhelm University in Bonn under the heading "Rather Meaningless Results". It reads as follows (translated by the author):

Are these really breathtaking new understandings of science with relevance to its predictions? My simple answer is: no. These are certainly results of expensive space experiments, requiring a great deal of effort, but they are rather meaningless, for example, with regard to the fate of our planet Earth over the next 100 years.

Because in the meantime, nearly 50 years after their "invention" by Stephen Hawking and his mentor Roger Penrose and their official announcement by John A. Wheeler, we have noted that not a single phenomenon exists in the sky, from which their existence could be reliably derived. Recently, even more drastically, black holes are supposed to be accreting, i.e., swallowing everything coming their way and gathering mass and attractive force so fast that one of them would surely already have swallowed our Solar System in the meantime. We should thus find ourselves in its interior, torn apart by tidal forces and burnt. In other words, we may conclude that they do not exist. They were a premature explanation. So why does the young generation not care about new insights?

> Black hole research pays well and the above insight has as yet not been captured by Google. Max Planck once pointed out that unrealistic scientific ideas need two generations to die out: the generation of their proponents and the generation of their pupils. I do not know a single phenomenon in the sky which cannot be better described or understood without black holes. But without them our branch of science would be significantly more boring.

According to Kundt, it seems that we are back to square one. But Kundt is a lone, dissenting voice among his peers. However, considering the unsolved riddles mentioned earlier, and especially the continuing lack of a GUT or even of a TOE, it seems that more than ever before, someone is needed to bring order to things that are once again falling apart—maybe someone like Johannes Kepler.

Johannes Kepler's Complete Works

The Bavarian Academy of Science is working on a historical critical edition of the complete works of Johannes Kepler [33]. The following list contains the present contents (2017) and gives some idea of Kepler's prolific publications.

Vol. I: Mysterium Cosmographicum. De Stella Nova. Ed. Max Caspar. 1938. 2. unaltered edition 1993. XV
contains: Mysterium Cosmographicum (1596) 3–80/Joachim Rheticus: Narratio prima, una cum Encomio Borussiae scripta (1596) 81–145/De Stella Nova (1606) 149–356/De Iesu Christi vero anno natalitio (1606) 357–390/Gründtlicher Bericht von einem Newen Stern (1604) 393–399
Vol. II: Astronomiae Pars Optica. Ed. Franz Hammer. 1939
Vol. III: Astronomia Nova. Ed. Max Caspar. 1937. 2. Unaltered edition 1990
Vol. IV: Minor publications 1602–1611. Dioptrice. Ed. Max Caspar and Franz Hammer. 1941
contains: De Fundamentis Astrologiae certioribus (1601) 7–35/De Solis Deliquio (1605) 39–53/Außführlicher Bericht Von dem diß 1607. Jahrs erschienenen Cometen (1608) 57–76/Phaenomenon singulare seu Mercurius in Sole (1609) 79–98/Antwort Auff Röslini Discurs (1609) 101–144/Tertius Interveniens (1610) 147–258/Strena seu De Nive Sexangula (1611) 261–280/Dissertatio cum Nuncio Sidereo (1610) 283–311/Narratio de observatis a se quatuor Iovis satellitibus (1611) 315–325/Dioptrice (1611) 329–414
Vol. V: Chronological publications. Ed. Franz Hammer. 1953
contains: De vero anno natali Christi (1614) 7–126/Widerholter Außführlicher Teutscher Bericht [vom Geburtsjahr Christi] (1613) 129–201/Ad Epistolam Sethi Calvisii Chronologi Responsio (1614) 205–217/Eclogae Chronicae (1615) 221–370/Kanones Pueriles (1620) 373–394
Vol. VI: Harmonice Mundi. Ed. Max Caspar. 1940
contains: Harmonices Mundi Libri V (1619) 7–377/Apologia adversus Demonstrationem Analyticam Roberti de Fluctibus (1622) 381–457
Vol. VII: Epitome Astronomiae Copernicanae. Hrsg. von Max Caspar. 1953. 617 S., 178 Abb.

© Springer Nature Switzerland AG 2020

W. Osterhage, *Johannes Kepler*, Springer Biographies,
https://doi.org/10.1007/978-3-030-46858-3

Vol. VIII: Mysterium Cosmographicum. Editio altera cum notis. De Cometis. Hyperaspistes. Edited by Franz Hammer. 1955
contains: Mysterium Cosmographicum (Editio altera) (1621) 7–128/De Cometis libelli tres (1619–1620) 131–262/Tychonis Brahei Dani Hyperaspistes (1625) 265–437
Vol. IX: Mathematical publications. Ed. Franz Hammer. 1955
contains: Nova Stereometria Doliorum Vinariorum (1615) 7–133/Messekunst Archimedis (1616) 137–274/Chilias Logarithmorum (1624) 277–352/Supplementum Chiliadis Logarithmorum (1625) 353–426
Vol. X: Tabulae Rudolphinae. Edited by Franz Hammer. 1969
Vol. XI,1: Ephemerides novae motuum coelestium. Edited by Volker Bialas. 1983
contains: Ephemerides Novae Motuum Coelestium ab anno 1617[–1620] (1617/1619) 7–134/Ephemeridum Pars Secunda ab anno 1621 ad 1628 (1630) 135–300/Ephemeridum Pars Tertia a 1629 in 1636 (1630) 301–458/Jakob Bartsch Offener Brief an Johannes Kepler (1628) 461–465/Ad Epistolam Bartschii Responsio (1629) 467–473/De raris mirisque Anni 1631 Phaenomenis (1629) 475–482
Vol. XI,2: Calendaria et Prognostica, Astronomica minora. Somnium. Edited Volker Bialas and Helmuth Grössing. 1993
contains: Practica Auff 1597 (1596) 7–17/SchreibCalender Auff 1598 (1597) 19–44/Practica Auff 1599 (1598) 45–55/Collectanea ad Prognosticum Anno 1600 59/Calendarium und Bericht vom feurigen Triangel 1603 (1602) 61–79/Prognosticum auf 1604 (1603) 81–100/Prognosticum auf 1605 (1604) 101–123/Prognosticum auf 1606 (1605) 125–135/SchreibCalender Auff 1618 (1617) 137–153/Prognosticum auf 1618 (1617) 155–172/Prognosticon auf das 1618. und 1619. Jahr (1618) 173–188/Prognosticum auf 1620 (1619) 189–215/Discurs Von der Grossen Conjunction 1623 Sambt Prognostico (1623) 217–245/Prognosticum auf 1624 (1623) 247–264/Appendix ad Progymnasmatum Tomum Primum (1602) 269–272/Astronomischer Bericht Von Zweyen im 1620. Jahr gesehenen Mondsfinsternussen (1621) 275–293/Terrentii Epistolium Cum Commentatiuncula Joannis Kepleri (1630) 297–314/Somnium Seu Opus posthumum de Astronomia Lunari (1634) 317–438
Vol. XII: Theologica. Witch Trial. Tacitus Translation. Poems. Edited by Jürgen Hübner, Helmuth Grössing, Friederike Boockmann and Friedrich Seck. Editorial Office: Volker Bialas. 1990.
contains: De Omnipraesentia Christi 7/Unterricht Vom H. Sacrament (1617) 11–18/Glaubensbekandtnus (1623) 21–38/Notae ad Epistolam Matthiae Hafenrefferi (1625) 39–62/Conclusion Schrifft [Hexenprozeß gegen Katharina Kepler] (1621) 65–100/Taciti Historische Beschreibung. Das Erste Buch (1625) 103–175/Gedichte 177–265
Vol. XIII: Letters 1590–1599. Ed. Max Caspar. 1945
Vol. XIV: Letters 1599–1603. Ed. Max Caspar. 1949. 2. unaltered edition 2001
Vol. XV: Letters 1604–1607. Ed. Max Caspar. 1951. 2. unaltered edition 1995
Vol. XVI: Letters 1607–1611. Ed. Max Caspar. 1954
Vol. XVII: Letters 1612–1620. Ed. Max Caspar. 1955

Vol. XVIII: Letters 1620–1630. Ed. Max Caspar. 1959
Vol. XIX: Documents concerning Kepler's life and work. Edited by Martha List. 1975
Vol. XX,1: Manuscripta Astronomica (I). Edited by Volker Bialas and Friederike Boockmann. 1988
contains: Apologia Tychonis contra Ursum scripta 15–62/Judicium de hypothesibus Tychonianis 65–82/Refutatio libelli, cui titulus Capnuraniae Restinctio 85–87/Catalogus librorum a Tychone Brahe scriptorum [confectus] 91–95/Problemata astronomica 99–144/De motu terrae 147–180/Aristotelis Buch von der oberen Welt. [13. und 14. Kapitel] 150–167/Responsio ad Ingoli Disputationem 168–180/Hipparchus 183–268/Lunaria 271–320/Restitutionum lunarium adversaria 321–392/Consideratio observationum Regiomontani et Waltheri 395–455
Vol. XX,2: Manuscripta Astronomica (II). Commentaria in Theoriam Martis. Edited by Volker Bialas and F. Boockmann/J. Kuric/I. Noeggerath. 1998
contains: preliminary work for Astronomia Nova
Vol. XXI,1: Manuscripta astronomica (III). De calendario Gregoriano. Manuscripta mathematica. Edited by Volker Bialas, Friederike Boockmann, Eberhard Knobloch, Hella Kothmann, Johanna Kuric, Hans Wieland. 2002
contains: Eclipses Lunae et Solis 9–212/Vorarbeiten zum Mysterium Cosmographicum 215–242/Vorarbeiten zu Astronomiae Pars Optica 243–261/Vorarbeiten zu De Stella Nova 263–290/Ergänzungen zu Commentaria in theoriam Martis 291–308/Vorarbeiten zu Narratio de observatis a se quatuor Iovis satellitibus 309–312/Vorarbeiten zu Epitome Astronomiae Copernicanae 313–329/Vorarbeiten zu Tabulae Rudolphinae 331–345/De Calendario Gregoriano (lateinische wie deutsche Schriften) 349–439/De quantitatibus libelli 445–461/De genesi magnitudinum 462–480/Manuscripta Arithmetica 481–519/Manuscripta Geometrica 521–590
Vol. XXI,2.1: Manuscripta harmonica. Manuscripta chronologica. Edited by Volker Bialas and Friedrich Seck. 2009
contains: De stellis 7–15/Brevis et dilucida explicatio fundamentorum harmonicorum 16–19/Ex dialogis Vincentii Galilaei De musica 20–31/Appendix ad Harmonices Mundi librum V (Keplers Kommentar zur Harmonik des Ptolemäus) 32–108/In libellum Sleidani de IV monarchiis 111–145/De septuaginta hebdomadibus in Daniele discursus 146–178/Chronologia a mundo condito 179–444/Astronomia Chronologiae Salutem plurimam 445–458/Dispositio historicorum in chronologia 459–461
Vol. XXI,2.2: Manuscripta astrologica. Manuscripta pneumatica. Edited by Friederike Boockmann, Daniel A. Di Liscia, Daniel von Matuschka and Hans Wieland. 2009
contains: Kepler's Horoscope collection

Timeline

Table A.1 gives a timeline encompassing events from the history of science or natural history relevant to our subject and treated here. Included are detailed incidents from the life of Johannes Kepler. The timeline is not to scale.

Table A.1 Timeline

570 BC	Birth of Pythagoras
490 BC	Birth of Anaximander
384 BC	Birth of Aristotle
365 BC	Birth of Euclid
165	The Canon of the Holy Scriptures
1401	Birth of Nikolaus Cusanus
1473	Birth of Copernicus
1483	Birth of Martin Luther
1492	Discovery of America
1509	Birth of John Calvin
1533	Birth of Elizabeth I
1543	Publication of "De revolutionibus orbium coelestium"
1545	Beginning of the Counter-Reformation
1546	Birth of Tycho Brahe
1547	Birth of Miguel Cervantes
1548	Birth of Giordano Bruno
1550	Birth of Michael Maestlin
1562	Beginning of the French Wars of Religion
1564	Birth of Galileo in Pisa
	Birth of William Shakespeare

(continued)

© Springer Nature Switzerland AG 2020

W. Osterhage, *Johannes Kepler*, Springer Biographies,
https://doi.org/10.1007/978-3-030-46858-3

Table A.1 (continued)

1571	Birth of Johannes Kepler
1575	Kepler takes ill with smallpox
1579	Kepler relocates with his parents to Ellmendingen
1583	First colony in Canada
	Kepler passes the "Landexamen"
1586	Kepler begins studies in Maulbronn
1588	Defeat of the Spanish Armada
1589	Kepler begins his studies in theology in Tubingen
1591	Kepler obtains his Masters degree
1591	Kepler takes up his position as mathematician in Graz
1595	Kepler publishes his "Mysterium cosmographicum"
1596	Birth of René Descartes
1597	Kepler marries Barbara Mueller
1598	End of the French Wars of Religion
1599	Invitation by Tycho Brahe for Kepler to work with him in Prague
1600	Execution of Giordano Bruno
	Meeting of Kepler and Tycho Brahe; Kepler takes up his position in Prague
1601	Death of Tycho Brahe; Kepler becomes Court Mathematician
1602	Birth of Kepler's daughter Susanna
1604	Appearance of a supernova
	Birth of Kepler's son Friedrich
1606	Publication of Kepler's "Astronomia Nova"
1607	Birth of Kepler's son Ludwig
1608	Lippershey invents a telescope
1610	Galileo detects Jupiter's satellites
1611 1612 1613 1615	Kepler describes his telescope in "Dioptice" Death of Emperor Rudolf II Kepler takes up a position as mathematician in Linz Kepler makes the acquaintance of Matthias Bernegger Kepler marries Susanne Reuttinger Beginning of the witch trial against Kepler's mother
1618	Beginning of the Thirty Years War
1619 1621	Publication of "Joannis Kepleri Harmonices Mundi Libri Quinque" Release of Kepler's mother
1625 1626 1627	New Amsterdam Kepler travels to Ulm Kepler enters the services of Wallenstein Publication of the Tabulae Rudolphinae
1629 1630	Birth of Christiaan Huygens Kepler's death

(continued)

Table A.1 (continued)

1631	Death of Michael Maestlin
1633	Galileo's renunciation before the Inquisition
1634	Wallenstein's death
1642	Birth of Isaac Newton
	Death of Galileo
1648	End of the Thirty Years War
1858	Birth of Max Planck
1879	Birth of Albert Einstein
1881	Michelson experiment
1889	Birth of Edwin Hubble
1900	Foundation of quantum physics by Max Planck
1905	Special theory of relativity
1915	General theory of relativity
1942	Birth of Stephen Hawking
1971	Space probes Mars 2 and 3 reach the planet Mars
1990	Hubble Telescope placed in orbit
2009	Kepler Telescope placed in orbit
2018	Death of Stephen Hawking

References

1. C. F. V. Weizsaecker: “Grosse Physiker“, Hanser, Munich, 1999.
2. M. Foucault: “The Order of Things“, Vintage Books, New York, 1973.
3. W. Oberschelp: “Bahnberechnung und Komputistik als Erkenntnisquellen in der Geschichte der Astronomie“, Informatik Spektrum, Bd. 32, Heft 6, December 2009.
4. A. Einstein, H. u. M. Born: “Briefwechsel 1916–1955”, Nymphenburger Verlagshandlung, Munich, 1969.
5. M. Gessen: “Perfect Rigor”, Houghton Mifflin Harcourt, Boston/New York, 2009.
6. J. Magueijo: “A Brilliant Darkness”, Basic Books, New York, 2009.
7. M. Caspar, Ed.: “Mysterium Cosmographicum” (Translation), Filser, Augsburg, 1923.
8. E. Lohrmann: “Hochenergiephysik”, Teubner, Stuttgart, 2005.
9. J. Mansfeld: “Die Vorsokratiker”, © Reclam, Stuttgart, 2008.
10. H. u. W. Hemminger: “Jenseits der Weltbilder”, Quell, Stuttgart, 1991.
11. M. Carrier: “Werte in der Wissenschaft“, Spektrum der Wissenschaft, February 2001.
12. Plato: “The Allegory of the Cave”, Republic, VII, 514 a,2 to 517 a,7, Translation by Thomas Sheehan, https://web.stanford.edu/class/ihum40/cave.
13. John Brockmann: “Einstein, Frankenstein & Co”, Scherz, Bern, 1990.
14. Carl F. v. Weizsaecker: “Die Geschichte der Natur”, Vandenhoeck & Ruprecht, Gottingen, 1954.
15. Dietmar Herrmann: “Die antike Mathematik”, Springer Spektrum, Heidelberg, 2014.
16. Wolfgang Osterhage: “Galileo Galilei—At the Threshold of the Scientific Age”, Springer International, Cham, 2018.
17. H. Bohn “Leitfaden der Physik”, Leipzig, 1915.
18. F. Krafft (Ed.) “Johannes Kepler, Astronomia nova—neue, ursächlich begründete Astronomie”, Wiesbaden, 2005.
19. P. Galluzzi “The Lynx and the Telescope“, Leiden, 2017.
20. G. Roeschert “Ethik und Mathematik”, Stuttgart, 1985.
21. H. Meschkowski (Ed.), “Moderne Mathematik”, Munich, 1991.
22. W. Osterhage, “Mathematical Theory of Advanced Computing”, Heidelberg, 2019.
23. D. Wiederkehr, “In den Dimensionen der Zeit”, Einsiedeln, 1968.
24. J. Huebner, “Die Theologie Johannes Keplers zwischen Orthodoxie und Naturwissenschaft”, Tuebingen, 1975.
25. H. Boettger, “Harmonices Mundi”, Magdeburg, 2005.
26. G. Mann, “Wallenstein”, Frankfurt am Main, 1971.
27. F. Broockmann et al., “Nicht das Kindt mit dem Badt außschütten”, Akademie aktuell 04, Munich, 2008.
28. http://www.hrad-cheb.cz/de/krvava-hostina.
29. D. Hattrup: “Die Wirklichkeitsfalle”, Herder, Freiburg, 2003.

© Springer Nature Switzerland AG 2020

W. Osterhage, *Johannes Kepler*, Springer Biographies,
https://doi.org/10.1007/978-3-030-46858-3

30. W. Osterhage: "Geometric Unification of Classical Gravitational and Electromagnetic Interaction in Five Dimensions: a Modified Approach", Z. Naturforsch. 35 a, 302–307 (1980).
31. C. Hartfeldt et al.: "Mathematik in der Welt der Töne", Magdeburg, 2002.
32. W. Fucks: "Mathematische Musikanalyse und Randomfolgen, Musik und Zufall", Gravesaner Blaetter, Vol. VI, No. 23/34, 1962.
33. https://kepler.badw.de/kepler-digital.html.

Name Index

© Springer Nature Switzerland AG 2020

W. Osterhage, *Johannes Kepler*, Springer Biographies,
https://doi.org/10.1007/978-3-030-46858-3

Subject Index

© Springer Nature Switzerland AG 2020
W. Osterhage, *Johannes Kepler*, Springer Biographies,
https://doi.org/10.1007/978-3-030-46858-3

V

W

Printed in the United States
by Baker & Taylor Publisher Services